I0033344

THE BEST AUSTRALIAN SCIENCE WRITING 2025

ZOE KEAN is an award-winning science writer with a focus on evolution, ecology and the environment. She has published in the *Guardian*, the ABC online, *The Best Australian Science Writing 2022, 2023* and *2024*, *Cosmos Magazine* and with the BBC. Her book *Why Are We Like This? An evolutionary search for answers to life's big questions* was released in 2024.

TEGAN TAYLOR is a multi-award-winning health and science reporter for the ABC. She hosts Radio National's *Life Matters* and the hilarious health podcast *What's That Rash?*

THE BEST AUSTRALIAN SCIENCE WRITING 2025

EDITORS
ZOE KEAN AND
TEGAN TAYLOR

FOREWORD
VEENA SAHAJWALLA

NEWSOUTH

UNSW Press acknowledges the Bedegal people, the Traditional Owners of the unceded territory on which the Randwick and Kensington campuses of UNSW are situated, and recognises the continuing connection to Country and culture. We pay our respects to Bedegal Elders past and present.

A NewSouth book

Published by
NewSouth Publishing
University of New South Wales Press Ltd
University of New South Wales
Sydney NSW 2052
AUSTRALIA
https://unsw.press/

Our authorised representative in the EU for product safety is Mare Nostrum Group B.V., Mauritskade 21D, 1091 GC Amsterdam, The Netherlands (gpsr@mare-nostrum.co.uk).

Introduction © Zoe Kean and Tegan Taylor 2025
First published 2025

A catalogue record for this book is available from the National Library of Australia

ISBN   9781761170508 (paperback)
        9781761179259 (ebook)
        9781761178566 (ePDF)

Cover design George Saad
Design Josephine Pajor-Markus

UNSW
SYDNEY

C©PYRIGHTAGENCY
CULTURAL FUND

CONTENTS

FOREWORD: THE NEED FOR A SUSTAINABLE FUTURE

Professor Veena Sahajwalla

Climate and energy concerns present humanity with an existential threat, demanding a radical shift in our relationship with the planet and how we think about its resources. Addressing this climate and energy challenge necessitates a multifaceted approach that leverages the power of scientific innovation, particularly where engineering, materials science and sustainability come together. Scientific research is how we will solve our big human challenges, and the stories within this publication show researchers striving to address the biggest of them all in our lifetime.

A sense of the urgency involved comes through in this anthology: from the young scientist looking out at Arctic ice sheets and considering that they may no longer form by the end of her career, to a detailed look at nuclear fusion research in China.

Waste as a resource

Australia's waste and resource recovery industry is increasingly challenged by complex wastes, such as electronic or e-waste and batteries. The Australian Bureau of Statistics' latest waste estimate figures show the Australian economy domestically generated 539 000 tonnes of e-waste in 2019, with more than 50 per cent going to landfill. Only 17.4 per cent is said to be recycled, but much of this goes offshore, where outcomes are unknown. Traditional Australian recycling facilities are set up to separate, dismantle or shred only, not to isolate the valuable metal alloys, rare earth elements and critical metals contained in e-waste.

In my own work, I grapple with the rising tide of waste and the global inability to recycle and reform the valuable materials contained in so many waste streams, at any scale. To mitigate these impacts and safeguard the future of our planet, we must transition to a sustainable, low-carbon economy where waste provides raw materials for remanufacturing.

Materials science and engineering can play a pivotal role in this transformation, with their focus on translating scientific discoveries into practical solutions. But we have a long way to go develop the innovations required to provide more sustainable, scalable solutions.

One story in this compendium that resonated with me concerns an innovative response to the problem of 'ghost nets' – huge discarded fishing nets which pervade the seas and oceans, causing great harm to marine environments and creatures. The non-profit Tangaroa Blue re-uses old, unwanted GPS-enabled buoys to keep track of the nets cut loose or lost in the Gulf of Carpentaria – using trash to track trash. My own UNSW Sustainable Materials Research and Technology (SMaRT) Centre also looked at the problem of these ghost nets, and found one of our technologies could be used in remote locations to reform this plastic waste either directly into new products or into raw materials for remanufacturing, creating localised circular economies.

The need for innovation

Dealing with waste in sustainable ways requires us to challenge conventional wisdom and explore unconventional approaches. This has been the basis of the work we do at the SMaRT Centre at the University of New South Wales, where we pioneer the science of microrecycling and develop technology-based solutions to reform hard-to-recycle waste streams into value-added products and raw materials for remanufacturing.

Because we understand these wastes at the molecular level, we can sustainably recover and reform valuable materials from no- or low-value wastes and help to truly create a circular economy and a more sustainable future. Our best-known innovations are Green Steel Polymer Injection Technology™, which uses waste rubber tyres as a partial replacement for the coke and coal needed in electric arc furnace steel-making, as well as our various modular MICROfactorie™ Technologies that use different waste types to create products and raw materials for remanufacturing. Our Plastics module, for instance, reforms hard plastics from e-waste into filament for 3D printing.

The intersection of engineering, materials science and sustainability is a fertile ground for innovation. Engineering provides the tools and techniques to design and build sustainable systems, while materials science offers insights into the properties and behaviour of materials. Together, they can create solutions that are both effective and sustainable – as is demonstrated by many of the stories in this publication.

Addressing the climate emergency also requires creative problem-solving. Traditional approaches may not be sufficient to tackle the complex and interconnected challenges we face – scientists must explore unconventional solutions and interdisciplinary approaches. One example of creative problem-solving in this area is the development of bioengineered materials. Researchers are creating materials that mimic natural processes, such as photosynthesis, to capture and store carbon dioxide. These materials can be used in construction, reducing the carbon footprint of buildings. Similarly, bioengineered plants can be designed to absorb pollutants from the air and soil, helping to restore ecosystems damaged by industrial activities.

A diverse scientific community is crucial

A diverse and inclusive scientific community is essential for driving innovation and ensuring that sustainability solutions are equitable and just. Diverse perspectives, backgrounds and experiences bring a richness to scientific inquiry and lead to more robust and creative solutions. By fostering a culture of inclusion and providing equal opportunities for all, we can harness the full potential of the scientific community to address the challenges of sustainability.

And this benefit is clearly borne out through the wide range of perspectives and ideas covered in this book. The diversity of writers and subject matter is pure inspiration, and demonstrates novel approaches to both understanding and interrogating the challenges faced around the world.

The challenges we face are significant, but by fostering inter-disciplinary collaboration, embracing future-focused technologies, and cultivating a diverse and inclusive scientific community, we can develop the innovative solutions needed to translate science into applications that help create a sustainable future for all.

Scientific discovery and science writing itself can create greater connectivity between and across markets, sectors, communities and cultures, enabling pathways to achieve new national and global opportunities. I commend this anthology, *The Best Australian Science Writing 2025*, as we rethink our approach for a sustainable future.

INTRODUCTION: A CONSTELLATION OF BRIGHT MINDS

Zoe Kean and Tegan Taylor

Gaze into a telescope and what do you see? Well, if it's a cloudless night and the scope is adjusted just right, you will see some of the planets that make up our solar system. Venus shining bright, Mars a little red, Saturn with its rings and ... is that yellowish one Jupiter? Before long you're adjusting the thing again and again, getting lost in the splendour of the stars. They congregate in a large streak across the sky: the Milky Way. This is our home galaxy. Just as the planets we spot are part of our solar system, together these stars, including our Sun, make up the swirling celestial community that appears in our skies as a splash of tiny white points – on a clear night, more than we could count.

And science? Well, the big, beautiful picture of science is made up of numerous tiny, twinkling data points. If we zoom in on a planet, it starts to become familiar: 'Yes! Definitely Jupiter, I can see its bands and big spot!' In the same way, zooming into this sea of data points, we find familiarity. We find the story behind the data. The stories in this book do that, joining the dots to tell tales of the universe that scientists have worked hard to reveal. The pieces range from the sublime to the silly, the mournful to the majestic. They encompass hope, fear and the type of awe you feel looking through a telescope on a crisp night.

Each story in this book is its own data point, and together they paint a picture of science communication at a distinct moment in time. Many of the stories were published before a change in government in the United States destabilised much of the global research community. Datasets disappeared. Australian universities have had their international funding threatened. The idea of science as a source of truth has been questioned. In this light, some of the moments captured within this anthology feel almost quaint. There are tells in some works that indicate they were written in a time before the rules changed. This book, at this moment, provides an opportunity to reflect on where we have been, and where we are going.

Writers from all over Australia submitted stories for this anthology. Our inbox was brimming with fabulous pieces from inspiring writers, and for every story that is included in this anthology there are four or five worthy pieces that we weren't able to include. We are pleased to present writers that have never been featured in this anthology before. However, keen followers of *The Best Australian Science Writing* will recognise many contributors' names. As editors, we have mixed feelings about this. Of course, we celebrate it: we are proud to showcase the work of a talented cohort of dedicated writers who have managed to forge long careers in science writing. However, the pool of working science writers in Australia is smaller than it should be – and shrinking. Media companies are under pressure, and their tightening budgets mean less money for staff – including specialist science reporters. The loss means science news is often covered by reporters with minimal background knowledge, with the risk that important nuance may be lost.

That's not to say journalism is the only genre of science writing, and we're proud to present an anthology that includes perspectives from scientists both established and early career; poetry; essays; book excerpts; and stories that defy categorisation ... along with investigations and reportage.

We're also grateful to see diversity in many forms among our contributors. However, a lack of submissions coming directly from

First Nations writers makes us wonder if there's more Australia could be doing to invest in, nurture and platform Indigenous science writing.

The stories within speak for themselves, but standing back reveals a picture of the world in 2024–25, and threads of commonality across various fields.

In April 2024, the world was sweating through its tenth warmest month on record. Alarmingly, the recorded temperatures were higher than researchers had modelled. In a political atmosphere where climate data and models are often confusing and controversial, **Tyne Logan** expertly and clearly lays out the challenges scientists face when modelling our changing climate.

Every data point plugged into a climate model is hard won. **Mark Horstman** delightfully illustrates the windblown work of Antarctic researchers as they drill into floating ice shelves to suspend high-tech sensors beneath the freeze. While emphasising the grind of the expeditioners, Horstman reminds us of the researchers back in Australia eagerly welcoming and analysing the data their colleagues collect. On the other end of the planet, PhD candidate **Lucinda Duxbury** invites us along on the deck of a research vessel as it crashes through Arctic sea ice that in a few decades' time may fail to form.

The ecological stories in this anthology chronicle crises, but they are not without flashes of hope. Like a lizard jumping off a rock, **Anthea Batsakis**'s words leap off the page as she describes the joyful moment two scientists find a baby long-tailed skink. The species is data deficient – that is, not enough is known about it to assess the population's health. However, the scientists suspect the little lizards are on the edge of extinction. **Zowie Douglas-Kinghorn** questions how many species have gone extinct before they ever had a chance to be described by science. The muse of her wide-ranging and philosophical essay is the smooth handfish, which has only been described to science once. Sometimes data is collected, then lost. **Olivia Congdon**'s tale of a museum heist involves an eccentric crook and load of butterflies and reinforces how important specimens are as scientists scramble to describe the world through an extinction crisis.

Animals who have their behaviours, presence or bodies recorded as data have their own rich experience of the world – or at least we think they might. **Amalyah Hart**'s article surveys consciousness research, examining arguments for and against our six-legged friends' inner lives. **Drew Rooke** shows a similar curiosity in his piece examining the possibility of language in whales. He talks to some of the biggest fish in the whale research world about claims of chatting cetaceans ... as well as the scientists injecting a bit of caution into the topic. For an ocean creature less traditionally charismatic than whales, **Anne Casey**'s poetry contains a twist, opening up space for empathy.

Joy and absurdity burst through some of this year's stories. **Angus Dalton** uses luscious and lewd prose to describe the fervour that accompanies the opening of a corpse flower dubbed 'Putricia' at the Sydney Botanic Garden, as well as the science behind the stinky floral structure. **Belinda Smith** indulges our morbid curiosity, explaining why some venomous snakes can still bite after decapitation. **Jacinta Bowler** describes a scientist who used sophisticated carbon dating to determine the age of his own kidney stone – proving both the age of the stone and the fact that humans will forever be curious. Infants and toddlers make the perfect maverick scientists, testing out their environment and always asking 'why'. But **Donna Lu** investigates why our memories of this curious time are so sparse.

Fraught and fallible heroes loom large in the sciences, particularly in psychology. **Rachel Fieldhouse**'s remembrance of psychologist Philip Zimbardo wrestles with the tricky legacy of his infamous Stanford prison experiment, the trauma it caused and the scientist's unwillingness to acknowledge the darker side of his work. While the fame of some figures in science is scrutinised, many never got the recognition they deserved, at least in their lifetimes.

Alicia Sometimes' triptych of poems honours the work of three of these scientists. **Natasha May** delves into a timelier controversy, questioning whether the diagnosis of borderline personality disorder should be shelved, exploring the argument that this highly stigmatised condition is in fact post-traumatic stress disorder. The piece examines the role of gender in diagnosis and treatment, interweaving rigorous scientific debate with personal stories.

Peers and even our own bodies can let us down. **Angela Heathcote** illustrates how, for a survivor of a near-deadly animal bite, the response on social media can mean trauma comes in waves. In a generous and personal piece **Felicity Nelson** describes her and her partner's shared and years-long struggle with long COVID, and the delicate dance they need to observe as they slowly recover. And while science and medicine typically encourage distance and objectivity, **Michael Leach**'s poem centres kindness in a usually sterile setting.

Data itself may be dispassionate, but how it's used can be dystopian. For years research has suffered as human datasets are overwhelmingly collected from white and wealthy people, biasing medical care towards the highly sampled population. However, **Dyani Lewis** shows targeted sampling of ethnic minorities can be ethically fraught, and calls on academic journals to not turn a blind eye to coercive data collection, particularly among ethnic minorities in China. **Smriti Mallapaty** uses simple language to explain why it is so hard to ascertain a death toll for the ongoing war in Gaza. Both works show how the collection of data *or* its absence can inflict violence on already oppressed peoples.

The writing in this book is deeply Australian. But in the times in which we live, to talk about science in Australia is to contend with the political climate in the United States. In a piece published before the 2024 US election, **Jackson Ryan** comes face-to-face with one of his online tormentors, a climate denier from Texas … and finds they both have something to learn from each other. But what about the people in power? **Rich Haridy**'s essay explores the realm of psychedelic drugs as medicine, and to what extent the therapy aspect should be part of how they're measured and marketed.

On substances and taking the human body to its limits: some athletes will stop at nothing to win – including doping with derivatives of worm blood. **Matthew Ward Agius**'s delightfully gross piece will make you squirm *and* learn. But what if you test positive to a banned substance you've never taken? **Ellen Phiddian** explores the world of mirror molecules in a through-the-looking-glass journey both fascinating and unsettling. And speaking of unsettling, the uncanny valley of humanoid robots is the focus of **Owen Cumming**'s essay, asking why we're so obsessed with replicating ourselves. Still, our own bodies are sites of discovery in **Sally Montgomery**'s chest-squeezing exploration of her experience learning to freedive.

Most inclusions in the anthology are measured and calm, but **Linda McIver**'s piece, a speech she delivered to colleagues in tech, has a touch of fury bubbling within it. McIver, a computer scientist and data science educator ponders the limitations of generative AI and calls on her audience to shake up data education.

Robots, AI, spooky backwards chemistry ... if all this is enough to make you want to check yourself out of Earth entirely, you can do just that. **Ceridwen Dovey** explores the hospitableness of exoplanets, only to discover what it is that makes Earth so special.

So back to Earth, and to discover that, like datasets, when individual humans come together in community they can be most powerful. **Cat Williams** and **Jo Chandler**'s pieces are portraits of local communities empowered by science as well as their own culture, working together for their environment. Williams introduces us to the Kiwirrkurra rangers in the Western Desert, who have been hunting feral cats on Country for decades and are now teaming up with scientists in an effort to further understand the felines' impact on the ecosystem. Chandler takes on the contentious issue of carbon credits. With hopefulness, she respectfully portrays a community that has been able to live more sustainably and have better social outcomes as a result of selling carbon credits to not log their trees.

Because the threat of climate change is ever present. And while a government report on air conditioning in Australia may not seem like the most inspiring source material, **James Purtill**'s tracing of our zero-to-100 addiction to cooling systems is one of the more creative and clear pieces on climate we have read. While Purtill contends with Australia's energy situation, **Gemma Conroy** looks at China's energy future. Conroy's access to scientists working on the superpower's quest for nuclear fusion yields a piece that explains both the science and the politics of the chase with clarity and flair. And because Earth is a planet worth maintaining, why not use the problem to be part of the solution? At least, that's what the beach defenders **Clare Watson** has been talking to are doing in using trash to track other trash in the Gulf of Carpentaria.

But aren't so many of us drawn to science writing for the awe? The beauty of mathematics – and the contribution one particular concept has made to many fields of science – is explained with clarity in **Robyn Arianrhod**'s book excerpt. **Sara Webb** tells us that 'ripples in spacetime' are hard to spot. But in 2023 scientists announced they had done just that, meaning they had recorded gravitational waves for the first time – the result of over 100 years of scientific endeavour. Webb's explainer masterfully covers both the history and the science that led to this moment. Smaller waves that felt bigger because they were closer to home can be found in **Carl Smith** and **Peter de Kruijff**'s unravelling of a mysterious series of seismic waves that rang Earth like a bell. Stepping down from the intergalactic and even planetary level, **Fiona McMillan-Webster** shows us that awe can found at our back door, tracing the history of the evolution of plant life through all that's green and growing in the garden.

The universe is, of course, awe-inspiring. Mathematics is awe-inspiring. Deep time and evolution is awe-inspiring. But so are humans. And like all those data points that are gathered and analysed and synthesised and published, informing our understanding, so too do moments of connection between friends, families, lovers and even strangers add up to the story of our lives. But more than that – they contribute directly to science, as expertly demonstrated by **Tabitha Carvan** in her joyous and moving journey down the rabbit hole of PhD acknowledgements. Carvan's piece captures both the loneliness of science and the fact it rarely happens without community.

For millennia, humans have looked to the night sky and told stories from the stars. These stories come together as a constellation of sorts – forming the picture of a year in science writing. So here are all the sparkling parts that make up a whole: individual stories coming together to form an anthology that tells the story of a year in science, and individual scientists, writers and science writing enthusiasts (that's you, dear reader) who create the community around this book. Thank you for taking the time to pick this volume up and start reading. We think you're going to love it.

THE NIGHT I ACCIDENTALLY BECAME A CORPSE FLOWER'S BEDSIDE MANSERVANT

Angus Dalton

Putricia has reached the climax of her necrotic bloom when I arrive at her bedside at 10 pm. The corpse flower's glasshouse smells of a cadaver on the turn. And the woman who raised her, Alyse Baume, has stabbed the short, savage blade of a pruning shear into her back.

The first blooming of an endangered *Amorphophallus titanum* at the Royal Botanic Garden in 15 years has summoned thousands of Sydneysiders to this steamy greenhouse, like the buzzing flies and beetles hypnotised by Putricia's rot. She's spawned an online community of almost a million livestream viewers.

Now, the critical moment has arrived. Every flower's bloom is, after all, an invitation to procreate. It is time for Putricia's pollination – and it just so happens someone is needed to hold aside her plush velvet curtain.

I step forward into the glitz of spotlights and phone torches bathing Putricia. Reeling in her dizzying putrescence and squinting in the lights, I hold the curtain aloft and try not to breathe as Baume starts to saw.

She cuts a square the size of a drinks coaster in the back of Putricia's skirt. The sliced-out portions resemble watermelon rind. 'It's like a fairy door,' observes John Siemon, director of horticulture at the Royal Botanic Gardens.

The square hole has revealed a ring of hundreds of black stamen-like structures. These are Putricia's female flowers. Another ring of male flowers is hunched above; they will blossom hours later.

'Welcome, you've arrived at the perfect time,' an attendant purrs into a microphone as crowds of late-night viewers spill in. 'Blessed be the bloom.'

A small zip-locked bag of pollen is produced. Even though they possess male and female reproductive parts, corpse flowers rarely self-pollinate. This pollen was donated by a corpse flower that bloomed in Queensland and put on ice for this very moment.

The gardens' curator manager, Jarryd Kelly, clasps the bag in his hands, warming the pollen. He peels apart the ziplock and tips it upside down. Specks of ochre ambrosia fall into a petri dish.

The energy is jokey ('Anyone got a credit card?'), but there's a hot, tense undercurrent. We are Playing It Cool with dilated pupils. Hundreds of murmuring punters flood around the operation. The air is warm and wet, Sumatran-style. Everyone is sweaty. Top buttons pop.

Amid it all, Putricia sprawls on the lush plinth of her boudoir, fronds akimbo with all the sultry confidence of the most famous girl in this city. She's misted by steam, green spadix erect, anointed with fragrance distilled from the juice of a hundred festering bins.

Open on Siemon's phone is the corpse flower Kama Sutra: an academic paper called 'On the thermogenesis of the Titan arum'. The paper stipulates when Putricia will hit peak heat and putridity. Siemon has tracked a building heat spike, and it seems we are approaching the pocket of perfect fertility – which coincides with the full force of her stench.

The plumes come in waves, like Putricia is periodically pressing the bulbous pump of a cursed perfume. The revolting geysers are thrust forth by the thermogenic (heat-producing) action of her stewing core, which is approaching 30 degrees.

Just last year, Dartmouth College scientists analysed the fertile flesh of a corpse flower and discovered amino acids that serve as the precursor of a chemical called putrescine. This compound is produced in decomposing animals and is responsible for the stench of putrefying flesh.

Their conclusion: Putricia and her kin may mimic death on a molecular level.

Eyes are watering, and noses are pinched, but this moment must be seized. Two paint brushes serve as the instruments of pollination; one stubby, one attached to a slender black metre-long pole.

Baume douses the short paintbrush in the pollen and reaches through the square portal, gently stroking Putricia's sticky pollen-seeking stigmas. Then she goes in from the top with the longer pole, attempting to pollinate as many internal flowers as she can. (Baume is both a horticulturalist and a make-up enthusiast. She was born for this.)

This would normally be the job of a fly or beetle smothered in pollen. Pairs of pollinating insects, drunk on the scent of roadkill, often mate themselves within a corpse flower's spathe.

The paintbrushes withdraw. A smattering of applause. The deed is done.

Putricia has the capacity to produce 400 seeds – a number that rivals how many of her kin remain in the wild in her native Sumatra. Putricia's heat will soon spike again as her male flowers bloom, and the pollen will be gathered.

We mightn't know for weeks if the intimate pollination operation worked.

Before I let the velvet curtain fall, Putricia's attendants gaze at her for a moment and, cloaked in the stench of death, do as we always have: hope for new life.

✱ *When did your kidney stone start growing? ANTSO scientist carbon dated his to find out*, p. **38**
A museum heist 70 years ago is still causing a flutter in butterfly science today, p. **275**

HUMANOID ROBOTS: MADE IN OUR IMAGE

Owen Cumming

Humanoid robots have been a mainstay in literature and story-telling for centuries. Be they androids, automatons, cyborgs, droids or golems, we've been fascinated by the idea of beings made in our image, replicas of our own bodies and minds.

Talos, the mythical guardian of Crete, is one of the earliest stories we have about a mechanical being created with human form. Forged in bronze by the Greek god Hephaestus, Talos looked like a man, moved like a man, and could even think like a man. While the story dates back to 700 BCE, Talos was – as we would call him today – a robot.

This work of fiction is drawing ever closer to reality. Hiroshi Ishiguro's Geminoids, Sophia by Hanson Robotics, Optimus by Tesla … the prospect of a humanoid robot that can move, act, and think like a person is on the precipice of being realised. But that raises certain questions.

Do we want humanoid robots in our society, our workplaces, or maybe even our homes? And, if so, why?

A reflection of ourselves

Associate Professor Janie Busby Grant is psychology lead researcher at the University of Canberra's Collaborative Robotics Lab. Her research looks at the interactions between humans and robots, and she is keenly interested in our motivations for creating robots with human form.

'There are a number of different factors driving the development of humanoid robots, some of them psychological, and some of them practical,' says Busby Grant.

'Designing a robot that mimics our shape and capabilities means it should be able to travel through doors, stairs and rooms, operate cars and other equipment, and flick switches, turn knobs and operate interfaces the way we do.'

The world around us is designed for human bodies, so, on a practical level, a robot designed to share those environments could benefit from a human form. Of course, there is already a plethora of non-humanoid robots performing specific tasks in human environments – drones, articulated arms and cleaning robots – but none of them fulfil the wholistic replication of human ability that seems to be a goal of robotics.

There's more motivating this wave of humanoid innovation than just practicality, though. The act of creating something in our own image provides the opportunity for emotional connection and to have those feelings reflected back to us.

'Often designers are trying to encourage people to see the robot as "alive" and to attribute human-like features such as emotions, beliefs and preferences to it. Providing a humanoid body shape makes this easier for us to do and hopefully makes the robot more effective,' says Busby Grant.

Along with the potential practicalities of human-like design and the appeal of emotional connection, there's another reason we seem to anticipate the development of robots that will move and act like humans. We've been told, repeatedly, to expect them.

'Most people want humanoid robots to look and behave in a certain way, and fiction has been a key factor driving these ideas,' says Busby Grant. 'Companies are catering to social expectations. Movies, TV shows and fiction have all told us that robots will look roughly like us, and that technology will deliver us humanoid robots sooner rather than later.'

There's the additional narrative, bolstered by an epidemic

of loneliness, that robots will be sophisticated enough to act as surrogates for human companionship, or even intimacy. A society where a human persona can be substituted by a robot could face serious ramifications, but this, at least, may not be an immediate concern.

'There are serious social and ethical concerns around designing robots to replace or augment human companionship. But the systems we have currently, and in the near future, are not going to be able to replicate the complexity, nuance and positive outcomes of human engagement,' says Busby Grant.

It's hard to say exactly what roles humanoid robots may play. Mild-mannered butlers serving at our whim, intimate companions for a lonely population, or perhaps indominable mechatronic conquerors of humanity. Who's to say?

Humanoid robotics is fuelled by a dream to embody our ideals and fulfil our expectations, but a dream only takes you so far. Eventually, someone must work out how to actually do it, which, in this case, means building some of the most advanced machines to have ever existed.

Layers of technical complexity

'There's not one day that I go without thinking how amazing our body is,' says Professor of Robotics and Art Damith Herath, who founded and leads the Collaborative Robotics Lab at the University of Canberra.

For someone attempting to recreate the human form, every intricate detail is illuminated. Every sense, every structure, and every movement of the human body becomes an engineering marvel. But with decades of experience designing and building automated robots, Herath is painfully aware of how difficult it can be to create robots that look and move like their creators.

'We are trying to create humanoid forms that can work in the same environments as humans, but humans are complex beings,' says Herath. 'For example, there are about seven degrees of freedom just within the human arm. To expand that to some sort of humanoid form, you're talking about hundreds of degrees of freedom.'

One degree of freedom is the ability to move one joint in one direction, like bending and unbending your elbow. So, with somewhere between 250 and 350 joints in the human body, there's a lot for roboticists to consider.

'All of these different complexities need to work in tandem. You're looking at hundreds of motors and actuators, each one monitored through a central system. There are a huge number of different algorithms working to just keep the whole system from falling over,' says Herath. 'And that's without achieving any specific goal. If you want to get something done, that adds layers of mechanical, algorithmic and societal complexities as well.'

Even for tasks that humans might consider mundane, the complexity of the programming and engineering involved is baffling. Two PhD students on Herath's team are trying to 'grasp' how the human hand performs a task as simple as sorting rubbish.

'People can easily identify textiles or organic materials and pick them up without even thinking, right? But a plastic bag, versus a rotten fruit, versus a T-shirt – each requires a different kind of dexterity. We still don't have a robot that can actually do all of that in one go. It is such a such a complex problem,' says Herath.

But inherently, the human form implies that these robots are designed to work with and alongside humans. So, even once the mechanics have been worked out, what really makes a humanoid robot successful is their ability to interact with people.

The human touch

Professor Mari Velonaki is the director of the Creative Robotics Lab at the University of New South Wales, a research group specifically focused on the integration of robots into human society.

'Our aim is to produce technologies that enhance human experience,' says Velonaki.

With the rapid advancement of humanoid robotics, the Creative Robotics Lab helps to evaluate new technologies and design robots that are right for social implementation. When it comes to designing humanoid robots, one of the major challenges that Velonaki's team deal with is deciding how 'human' a robot needs to be.

'When we say humanoid robot, it's a very big category. The human references might be quite abstract, but you still can see something that appears like a head or the camera is like eyes. We can go from something very near realistic to something a bit more abstract. It's all humanoid robots,' says Velonaki.

Gauging what features a robot should have requires a deft touch. Human features can help us feel an affinity for machines, but some of the most lifelike humanoid robots, like Hiroshi Ishiguro's Geminoids, can become victims of the 'uncanny valley', the dreaded sensation of wrongness that comes from a robot seeming almost human, but not quite right.

'It's something that starts with attraction and turns to some sort of repulsion,' says Velonaki.

Making humanoid robots fit for human society isn't simply a matter of creating human replicas. Rather, the implementation of humanoid robots is about understanding what elements of humanity a robot needs.

'We need to think, "Why does this robot need to be humanoid?", and design for the situational context from the very beginning. Where is it going to be operating? What will it be doing? Who is interacting with it?,' says Velonaki.

The potential ways for humans and robots to interact are boundless, but it's not one-size-fits-all.

'Small, toy-sized humanoid robots might be helpful for young children on the spectrum to learn about social cues and social behaviours by playing. That's a good match. But for people in a nursing home, something that looks too much like a human can be highly problematic,' says Velonaki. 'Ageing populations can have sensory systems that might not be fully functional, and it creates an area of confusion: "Is that a human or not?" In that context, you need something that doesn't look threatening, and has some sort of human references, but is very clearly a robot.'

Ultimately, humanoid robotics is a science of interaction. Function is good: the robot needs to do what we want it to do. Trust is even better – something that we can look at with surety and comfort. Yet regardless of how well they might function, or how friendly they might look, for humanoid robots to really work in a human world, they need to have something that we want to interact with.

'There's nothing wrong with making something function as a matter of fact, but why don't we inject creativeness and playfulness into the objects or technologies that we use every day?' says Velonaki.

Humans are playful; it's in our nature. Playfulness is part of how we converse with the people and things that we care about, and creating that sensation can begin without even the need for language. When we speak to another person, perhaps 70 per cent of everything that passes between us is non-verbal. This tapestry of our expression is part of our humanity, informing others of our thoughts, moods, and intentions.

'If you're whispering, if you're singing, if you're moving, people know what you mean. These things are all cross-cultural and this is a great foundation that we can start building on,' says Velonaki.

We're hardwired to understand human expression, and even simple gestures can hold a depth of meaning. So, when we're designing ways to communicate with robots, why not use the signals that our brains already know?

'It's very important for emotional and mental health to be a little bit more playful with these technologies. An expression of a face is better than written text in terms of long-term cognitive load fatigue. It's much faster and easier to understand,' says Velonaki. 'If it's just a screen with a big smile, I don't have to go close, I don't have to put my glasses on, I don't have to read more text, I don't have to think about it!'

From Velonaki's perspective, the purpose of humanoid robots entering society isn't to create more work for people, or to replace the relationships that we already have. It is to make our lives easier and more enjoyable, taking on the jobs that we don't want so we can focus on what matters most.

'We want to be able to have humans do what humans do best, and that's connecting with people,' says Velonaki. 'It's holding Mrs Smith's hand and explaining to her what's going to happen before surgery, rather than destroying my lower back because I have to carry Mrs Smith from the bed to the wheelchair. I can see a robot in between, helping and making it entertaining, but always with a human in the loop.'

Guiding principles

Humanoid robots are only going to become more advanced as our technology develops. We need to prepare ourselves for radical changes in our societies, cultures and systems of commerce.

'Once AI is embodied in actual mechanical systems and they start to move around in the environment, that's going to disrupt a whole host of industries and socioeconomic stratums,' says Herath.

Roboticists and engineers are increasingly aware that their passionate drive for creation needs to be tempered by those who can provide perspective on how humanoid robots may affect society. As Herath puts it, quoting Professor of Electrical Engineering and Computer Sciences Jitendra Malik, 'Robotics is too important to be left in the hands of roboticists.' Roboticists need to work alongside sociologists, psychologists and ethicists to understand and scrutinise their potential impact.

'We need to be both realistic about what robots can do but also have the debates in coming decades about what we as society want them to do,' says Busby Grant.

The complexities of having robots with a human form, and some semblance of a human mind, entering our society won't be quickly resolved. We will need to reflect upon our own morals and motivations. How should a machine that is humanoid, but not human, be treated? Are they a practical solution to our problems, or just a fancy of fiction we've brought to reality, creating more problems than they solve?

A guiding principle when designing a technology that imitates humans could be to create something that brings more humanity into the world, rather than just more technology.

'Well designed robots don't make us like machines, they enhance our humanity. We want to design robots that remind us we are human,' says Velonaki.

✱ *Faster stronger higher doper*, p. **74**
  Insect consciousness, p. **87**

MATHEMATICAL TRANSFORMATIONS: THE POWER OF VECTORS AND TENSORS

Robyn Arianrhod

Every now and then, there's a spectacular breakthrough in our understanding of the world. The upending of the belief in our central place in the solar system, for example, and the relativistic revolution that changed the way we see time and space – and the way we see our place in the cosmos itself. The dramatic discovery of electromagnetic radio waves, which led to wireless technology and its wondrous transformation of our everyday lives – and the quantum revolution, with its seemingly magical new micro-technologies and its preternatural shattering of our notions of 'reality'. And then, of course, there's the digital revolution that is still changing the way we communicate with each other, especially now that AI has become so sophisticated. These breakthroughs have brought into being new technological and cultural eras, and much has been written about them. It's less well known, however, that these scientific and technological paradigm shifts went hand in hand with equally dramatic mathematical revolutions. This is a story about some of those unsung revolutions ...

At its core, it's a story about the evolution of the way we humans record and make sense of all the data that swirl around us. In particular, it's about the dramatic mathematical transformations that gave us the remarkable concepts called 'vectors' and 'tensors', for they underlie so much of modern science – and much of our technology, too. They are languages that have helped us uncover mysteries of the universe as if we were gods.

The reason for this awesome power is that vectors and tensors made it possible to handle the dimensions of space in a new and transparent way – and this, in turn, made it possible to discover new laws of nature, and new technological applications of these laws. Anytime you want to pinpoint locations in space you need to handle these dimensions – rotating a robot arm, say, or designing a bridge or wind turbine; figuring out the effect of an electromagnetic force in a motor or generator, say, or predicting the path of an electromagnetic wave, a water wave, or even a gravitational wave; plotting the trajectory of a satellite or calibrating a guidance system such as GPS; or just about anything you can imagine doing in space or space-time …

But these languages are not just about physical dimensions – they're about 'dimensions' of information, too. You've likely read about 'big data' and the information revolution, but it is vectors and tensors that help make data usable and comprehensible – the way the periodic table of chemical elements is both an organisational and theoretical tool in chemistry, except that the maths in our story is so much more widely applicable.

Yet vectors and tensors themselves are remarkably simple – on the face of it, at least – for you can, indeed, begin by thinking of them simply as neat ways of representing information. For instance, you might remember from school that a vector can encode information about both the size and the direction of a physical quantity – say, a velocity or force. So, on a Cartesian coordinate grid with x and y axes (or x, y and z in three dimensions), you can represent the force or velocity with an arrow pointing in the required direction, while the arrow's length gives the size or 'magnitude.' Tensors add in more layers of information, so they are like multidimensional arrays rather than arrows. But when mathematicians discovered the rules for how these arrows and arrays combine with each other, they realised they had found a brand-new language for thinking brand-new thoughts. And this is a rather wonderful idea.

A simple illustration of what I mean by this is that for thousands of years mathematicians worked only with numbers. The evolution of real number systems was remarkable enough, but these numbers express only one thing: quantity – the magnitude of a weight, height, distance, amount of money, number of apples, and so on. Vectors and tensors, on the other hand, encode several things at once, which is why they are such great ways of representing a lot of data. And this extra information means that vectors and tensors can offer a far richer picture of an industrial or IT problem, say, or a physical model, than a single number ever could.

The first major physicist to recognise the power of vector language was the gently eccentric 19th-century Scottish laird James Clerk Maxwell. His theory of electromagnetism was the very first modern field theory, which he used to crack the long-standing riddle of the nature of light and to predict the existence of radio waves – all in one fell swoop. His initial theory was intuitively 'vectorial', but once he learned that vectors were actually 'a thing', with their own mathematical rules, he realised they were the right tools for expressing his discovery more succinctly and elegantly.

Not that many people took him seriously at first: his breakthrough application of 'vector fields' to represent nature's electromagnetic fields was ... well, just too mathematical, too 'unphysical', for mainstream physicists ...

As for tensors, Maxwell died before they were developed, but I'm betting he would have recognised their power, too. He died the same year that Einstein was born, which is especially symbolic – not only because Einstein's theories were inspired by Maxwell's but also because Einstein did for tensors what Maxwell had done for vectors: he was the first major physicist to show their practical power. They enabled him to create curved space-times and the discipline of modern cosmology – and they enabled him to predict the existence of gravitational waves and lenses, and to accurately quantify the gravitational effect on time that is now used to make GPS directions so accurate.

It took experimental physicists a quarter of a century to verify in the lab Maxwell's prediction of radio waves, and it took a hundred years to detect Einstein's gravitational waves. That's an indication of how far ahead of the game these vector- and tensor-based theories were. This kind of ability to make predictions is one of the exciting things about mathematical language. It's as if the act of describing physical reality mathematically creates a magnifying glass, revealing, through mathematical patterns, underlying physical attributes that had long lain hidden.

Since vectors and tensors are ways of storing and using information, they're useful far more widely than in physics alone, of course. As I intimated earlier, they are playing a fundamental role in a growing number of areas that need to handle a lot of data – from engineering and genetics to search engines and artificial intelligence, with much more in between.

For instance, in quantum mechanics, the axis of the magnetic orientation, or 'spin', of an electron can be 'up' or 'down' – or a 'superposition' of the two, as if the electron can't decide whether it wants its spin to be up or down. It's analogous to a two-dimensional velocity vector having one component in the direction x, say, and another in the direction y, so for electron spin you can let the two spin directions 'up' and 'down' be your axes.

Similarly, in computing you encode information with two binary digits ('bits'), 0 and 1, which are represented physically by, for example, turning an electric switch off and on. In quantum computing, the analogue of a bit is the qubit (pronounced 'cue-bit', short for quantum bit). Quantum 0's and 1's can be manifested physically as the two fundamental states of electron spin – 0 might be represented by spin 'up' and 1 by spin 'down' – so qubits, too, can be represented mathematically as two-dimensional vectors.

In other business and tech examples, the axes or 'dimensions' might represent words and documents in a search engine, or different questions on a website questionnaire or political survey, or the different factors affecting house prices and other socioeconomic data, or the positions and colours of pixels in image processing ... there are myriad possibilities!

Yet the development of the full power of these mathematical ideas was so astonishing, and so far-reaching, that I'm treating their discoveries as mathematical revolutions. It's helpful to think of tensors as a generalisation of vectors, but that is hindsight: it took 300 years to move from a fledgling form of vector language to a sophisticated language that incorporates vectors and tensors in a rigorous way. And to get to that first, nascent hint of the vector concept, it had already taken many centuries – millennia, in fact, if we go back to the oldest surviving mathematical records. For the history of vectors and tensors is linked with the history of the symbolic representation of data, and these ancient documents show that finding ways to represent information is at the heart of the story of mathematics itself ...

But when ancient Mesopotamian mathematicians etched tables of data into clay tablets, they surely never imagined the huge amount of data that tensors need to handle in physics and data science today. And when ancient Chinese mathematicians used matrix-like arrays to solve systems of linear equations, they would have had no idea how complex and diverse these systems are today – covering everything from optimising business costs and game strategies to robotics and AI, to search engines and more.

The 19th-century inventors of vector and tensor mathematics, too, had little idea how powerful these revolutionary languages would become, what new physical worlds and technological possibilities they would help open up. But once this new mathematics was out in the world, its patterns intrigued other mathematicians – especially those who like to see how far they can push the logical consequences of the rules and grammar of mathematics. And sometimes they push so far that they tumble into a new kind of reality, like Alice through the looking glass ... Such is the power of mathematics!

✱ *Sound of the slow-rolling sea*, p. **175**
The world has been its hottest on record for ten months straight.
Scientists can't fully explain why, p. **265**

SOLOMON ISLANDS TRIBES SELL CARBON CREDITS, NOT THEIR TREES

Jo Chandler

When head ranger Ikavy Pitatamae walks into the rainforest on Choiseul Island, the westernmost of the nearly 1000 islands that make up the South Pacific archipelago of Solomon Islands, he surveys it with the heart of a tribal landowner and the eye of a forester.

Leading the way up a track into the bush, he wades into a glassy stream, stirring small, brown fish into a spin. Surveys have identified some 50 freshwater species in these waters, a haven of biodiversity in a nation ravaged by high rates of logging. At the sound of a thumping whoosh overhead, Pitatamae points up just as two Papuan hornbills flash across a gap in the canopy. 'They always fly in pairs,' observes Wilko Bosma, a lanky Dutchman trailing behind the ranger. 'They're committed for life.'

Bosma made his own commitment to this forest after landing here 25 years ago as an idealistic forestry graduate, working alongside Indigenous tribal landowners on small, sustainable timber projects. That's how, in 2004, he came to establish with tribal partners a local conservation NGO – the Natural Resources Development Foundation (NRDF) – and got to know Linford Pitatamae, the older brother of ranger Ikavy and a leader of their tribe, the Sirebe.

Together with a handful of neighbouring tribes, the Sirebe resisted offers of quick money from the Malaysian companies whose ships lurk off the coast, piled with logs for export to China, and began laying plans to profit from allowing their trees to stand. Aerial maps reveal their remarkable success: the only expanse of green on the island that is not scarred by logging roads.

'If we misuse or destroy this land, we will not have any other,' explains Linford Pitatamae. 'So we have been committed to protect our lands, our forests, our rivers and streams, and all the resources for quite a long time.'

Finally, the Sirebe's long game is paying dividends. In 2019, the tribe's three-square-mile (777-hectare) forest became the first legally protected area in the nation, a 'beacon of conservation and natural resource management', according to Culwick Togamana, Solomon Islands' environment minister. In February 2022, the Sirebe became the nation's first landowners to receive payment from international investors for keeping their forests intact. With neighbouring tribes also securing protected areas and selling 'high integrity' carbon credits to buyers looking to offset their greenhouse emissions or burnish their green credentials, the Babatana Rainforest Conservation Project – named for the tribes' shared language group – now covers 26 square miles (over 6700 hectares) of protected tropical rainforest.

Ikavy Pitatamae opens his phone to record his sighting of the hornbills in a biodiversity app, then uploads a photograph of a fallen tree into another app used for tracking changes in forest conditions. The calculations underwriting the value of this project on the international carbon market began with painstaking surveys the rangers conducted over six months, prior to its verification in 2021. They measured everything growing and living within a list of survey sites 80 feet (24 metres) in diameter. Rangers repeat the surveys every ten years and annually conduct transect surveys that record every change and every animal spotted. Rangers routinely patrol boundaries, checking for incursions. Reports are verified by independent auditors and against satellite imagery.

Pushing through the dense tangle of vines and shrubs to the

base of a giant brown terminalia tree, Ikavy Pitatamae demonstrates how baseline data are collected. He passes a measuring tape around the tree's girth, then aims a laser beam up to where the trunk – likely 200 years in the making – branches out. Loggers would covet this specimen. Pitatamae's data points verify that this forest contains harvestable and marketable timber that would likely be felled were it not protected – thus demonstrating a core carbon market requirement that the trees would not otherwise exist without this project. The data are also necessary for tallying the forest's carbon store, or inventory. The magnificent canopy, the forest's messy middle storey of young trees and saplings, and the biomass of its understorey would all be just collateral damage to loggers coming for this tree's trunk. But left standing, they are part of the complex calculus paying the rangers' wages and supporting others in the tribal communities and villages.

Since 2022, the Sirebe have received five quarterly payments for their forest project. The money should keep flowing until 2045, when the contract comes up for renewal. The payments are made by Nakau, the Pacific-wide, rights-based, nonprofit operator that coordinates the Babatana project and works in collaboration with NRDF. Similar payments are in the pipeline for neighbouring tribes as their projects are verified and added to the Babatana project portfolio.

The Babatana project is owned by the tribes, who also retain their carbon rights. This structure, it is hoped, will help the project overcome some of the criticisms aimed at other carbon-trading schemes internationally, including that they perpetuate extractive colonial dynamics in their dealings with forest people and have led to episodes of conflict and violent dispossession. Many experts have been damning of various projects' lack of consultation with local people and transparency, and there is substantial evidence that some projects are pure greenwashing.

Among those critics is the Australian sociologist Kristen Lyons, a sustainability and development expert who spent years observing projects in East Africa that failed to deliver what they promised. With her fellow University of Queensland sociologist Peter Walters, she's taken the position that carbon offsets not only enable big industrial polluters to continue emitting, they also recruit traditional landowners 'as unwitting accomplices to this environmental procrastination', as the scholars wrote in a 2021 book chapter.

Nonetheless, after several years of monitoring the Solomons project, Lyons and Walters argue that it is a regionally significant example of best practice for carbon projects in a small nation. 'I have been, historically, incredibly critical of carbon offset projects,' Lyons says. But after applying an environmental justice lens to her study of the Babatana project, she notes, 'My view has shifted.'

Lyons observes that while forest owners should, in theory, be powerful stakeholders in a global market worth hundreds of billions of dollars, in reality they are often far from equal partners. Indigenous people may struggle with impenetrable jargon from the wrong side of the digital divide, often in their third or fourth language. But on Choiseul Island, local engagement is strong. Lyons says she has seen communities using money received from the offsets for programs that reflect their own priorities, including agricultural, sanitation and education projects, which will provide lasting value even if the carbon caper goes belly up tomorrow.

'We need to attend to human rights, particularly Indigenous rights, as we're seeking to attend to the climate crisis,' says Lyons. 'And so, if local communities – particularly Indigenous communities – are saying this is a pathway [they] want to explore, I think it would be a terrible thing if there was not the support [for it].'

Their work in the forest completed, the Pitatamae brothers and Bosma climb aboard their motorised canoe and slowly navigate down the shallows of the Kolombangara River, past crocodiles, kingfishers and flood-eroded banks. Big rains used to come once a season but now occur five or six times in succession, says Linford

Pitatamae. The damage is far worse in logged areas, where floods wash out crops and sediment contaminates clean water and coastal fishing grounds.

This river flows through 330 square miles (around 85 500 hectares) of lowland, riparian and montane forests that contain some of the richest biodiversity remaining in Solomon Islands. On the northern bank lies the 18-square-mile (4662-hectare) protected area of the Padezaka tribe, whose carbon project is in its last stages of verification. 'This area is heavily threatened by logging,' Bosma says. 'It's been quite an achievement that they could get it [protected] in time.' Some neighbouring tribes have let loggers in, and muck from the churned landscape, logging camps and roads is spilling into the water catchment.

Leaving the river, the canoe picks up speed for the last bumpy leg home, over open ocean, and finally pulls up on the sand at Sasamungga, at the midpoint of Choiseul's south coast. The sprawling village is home to around 1000 people from the Sirebe, Siporae, Vuri, Padezaka, Garasa, and Lukulombere tribes, whose ancestral forests are inland, in the river country we've just left. Like many Pacific forest people, they have over the decades drifted to hubs like Sasamungga to access education, markets, health services, transportation and jobs.

Only about 10 per cent of the small but growing Sirebe population have paid employment outside the village, Linford Pitatamae says, so most of the community have had a hand-to-mouth village life, relying on what they could earn from fishing and agriculture. Now, their carbon project sells around 17 423 credits a year on the international market under a deal extending 30 years. It takes in roughly $263 350 a year, of which 20 per cent goes to NRDF, 20 per cent to Nakau, the project coordinator, and 60 per cent to the 27 households of the project owners. In a subsistence economy in one of the poorest nations in the Pacific, that's life-changing income.

One man put his household's share into a new outboard motor and insulated boxes to store fish for sale. His neighbour invested in equipment for a mechanical workshop. Families have installed solar panels and toilets with septic tanks, upgraded their homes, and plumbed water taps that deliver rainwater collected in new tanks.

Five per cent of payments go into the Sirebe women's savings club – $2360 every quarter. Karah Qalo, who founded the club, says the 33 members use it to pay their children's school fees, buy materials for their food gardens, invest in bakery projects and bee-keeping, and run their phones. Their situation is radically different from that of women from forest communities that have been logged. Local custom largely holds that only men have authority over the land, while women have only the right to use the land – including for gardening, firewood or water collection. When male leaders allow loggers in, says Qalo, they get cash, but the women lose everything. 'The water will be polluted, air pollution, everything will not be good because of the pollution of the machines.'

Carbon marketeers usually describe the carbon that is sequestered and traded as the core benefit of their business and any positive social or economic outcomes for local people as co-benefits. But Nakau chief operating officer Alex McClean argues that from the perspective of forest people, 'the carbon is close to irrelevant. What matters to them are improvements to their life and livelihoods.' He likes to turn the measure of integrity on its head, so things like jobs for rangers and funds that empower women become core benefits, with carbon reduction the icing on the cake.

But carbon market pioneer Mark Trexler, while not commenting specifically on the Nakau model, questions this approach, and the application of an environmental justice lens to carbon projects. '"High integrity" in the offset space has to refer to climate change mitigation benefits,' he says, not to co-benefits including forest community impacts. 'If voluntary markets become seen primarily as a way to send money to impacted communities, we're talking about carbon contributions, not offsets.'

Nakau CEO Robbie Henderson says demand for the Babatana project is strong and that 'we could definitely sell our credits several times over at the moment'. Some buyers don't use those credits to offset anything, he says, but rather to demonstrate their environmental commitment. Some agree not to claim carbon neutrality; some offset residual emissions. An exclusion list prohibits the sale of credits to support any fossil fuel expansion.

Linford Pitatamae recalls plenty of scepticism when he and Wilko Bosma started talking to the community about the crazy notion of selling their forest, but not their trees. 'At first, we didn't think this was a real project,' Pitatamae says. 'The community didn't really understand.' It took years of 'continuous engagement, training [and raising] awareness' to win support, he says, then more years to work through the hurdles of verification and certification to get to this point. Now he's hearing from landowners across the island, and the nation, who are hungry to learn more.

'We believe we have to respect the environment – that will bring us a good future,' says Linford Pitatamae. 'We still have our trees [and] good drinking water sources. Our birds are still there. Our sacred places are still there.' He spends long hours most days locked away from the forest he loves, grinding through the paperwork and complexities of keeping this enterprise going. But he's not complaining. 'This is the right path for rural Solomons to follow,' he says.

✳   *Cat-astrophe: Australia's feral cat problem*, p. **33**
    Air conditioning quietly changed Australian life in just a few decades, p. **98**
    Using trash to track other trash, p. **209**

CAT-ASTROPHE: AUSTRALIA'S FERAL CAT PROBLEM

Cat Williams

The room is loud and full of chatter in Pintupi (pronounced Pintaby) language. Everyone is eager to share stories about cats. It's difficult to understand what's going on until Dannica interrupts and says in English, 'They're saying cats eat everything.'

I'm on a Zoom call with Dannica Shultz, who's dialling in from the Kiwirrkurra community, the most remote community in Australia, where she is the ranger coordinator.

The introduction of cats into Australia has a white history, but Traditional Owners, like the Kiwirrkurra rangers, are the best in the business of managing pest species. Yet feral cats cause more damage in Australia than anywhere else in the world.

Cats were on the ships when whitefellas first arrived in Australia in 1788. They were brought as pets, but later released to control rats and rabbits. Cats were introduced to Western Australia separately in the 1840s. 'They brang it [cats] from Europe,' says Jodie Ward, a Kiwirrkurra ranger. 'They been bringing feral animals for years.' Historical data and stories from Aboriginal people confirm that cats lived across the entire Australian continent by the start of the 20th century.

The *Illustrated Handbook of WA* from 1900 was one of the earliest references of cat meat as a food source for Aboriginal people. In 1935, a Western scientist noted that cats were the only non-Indigenous mammal granted an Aboriginal name – another indication of how long cats had been present in the landscape. All of these instances highlight researchers speaking with Aboriginal people to inform their scientific inquiry. Through their deep and ongoing connection with Country, they have much greater insight than Western scientists about the history and management of feral cats.

Oral histories point to the ongoing and increasing presence of feral cats. When I asked the Kiwirrkurra rangers how long cats had been on their Country, they all said that cats had been around for their whole lives.

An article for *Time Australia* magazine in 1994 reported that, when scientists spoke with Aboriginal people, they 'got a pretty consistent picture that ... cats had been present for a long time'. Another study from the late 1980s interviewed Aboriginal Elders from the central desert region, who all said cats had been present their whole life.

Modelling suggests that there are now between 1.4 million and 5.6 million feral cats across Australia, with an estimated 0.27 cats present per square kilometre.

Every morning when you wake up, the Australian population of feral cats has killed around 5.5 million animals while you were sleeping. Through discussions with Aboriginal people from the Ngaanyatjarra Lands in 1979, scientists from the then Department of Wildlife and Fisheries were able to create a database of which animals were present and which had been impacted by cat populations. Even though interviews with Aboriginal people were the centre of this study, none of them are mentioned by name.

Cats threaten over 100 native animal species and are the main contributor to Australia having the highest rate of mammal extinction in the world. Scott West, a Kiwirrkurra ranger, says 'Cats prey on birds, mammals and lizards.'

In Western Australia specifically, cats threaten 36 mammal, 22 bird and 11 reptile species and have caused the extinction of up to 27 mammals and ground-dwelling birds. After being declared a pest species in Western Australia in 2019, there was a national inquiry into feral cats in 2020, with six recommendations on how to reduce their impact. These centred mostly around increasing research and management strategies.

Combining Aboriginal knowledge and western science is the ultimate way forward to limiting the impact that feral cats can have on Western Australia's native animals. Kiwirrkurra's ranger program is leading the way in managing the feral cat population.

Kiwirrkurra is an Indigenous Protected Area (IPA), meaning that the 4.59 million hectares of land in and around Kiwirrkurra is managed strictly by Aboriginal people. The skills utilised by the Kiwirrkurra rangers to manage their land have been passed on from generation to generation.

They have a large team of rangers, including Jodie and Scott, as well as Yalti Napangarti, Yukultji Napangarti (Nolia Ward), Mantua James, Mary Butler, Kim West and Conway Gibson. The team works according to Kiwirrkurra's tjukurrpa, or Dreaming stories, using skills that have been passed on from generation to generation.

A subsection of this ranger team are the 'old ladies', made up of Yukultji, Yalti, Mantua and Kim. These women have been hunting cats since they were little kids and are considered master cat trackers. The old ladies can determine whether tracks are fresh, if the cat was walking or running, and what direction it was travelling in.

Cat hunts can be planned well in advance or be spontaneous if someone spots fresh cat tracks. Hunts tend to happen during the warmer months in the hottest part of the day and take anywhere from 30 minutes to a few hours. The hunts that are planned out start at a known threatened species site and have lots of people involved, including the kids so they can learn tjina (tracking).

Kiwirrkurra has ninu (bilbies) and tjalapa (great desert skinks), which are highly threatened by cats and are of great significance to the community. Yalti says that cats will 'wait at the burrow to eat the bilby'. Scott says that, when cats are brought back to community, they are a good food to share, because they provide more meat than goanna. 'It's better than shop meat,' says Jodie.

Cats are also believed to have medicinal purposes. Anecdotally, eating cat meat helps your heart, muscles, brains and eyes. However, caring for Country and a tasty meal aren't the only things motivating the cat hunters of Kiwirrkurra.

An integral part of cat management in the remote community is the bounty that hunters can claim. There's a $100 reward for killing a cat, but to claim the bounty money, Dannica says they have to 'bring the guts back to be frozen so we can analyse the stomach contents'.

At this point, someone excitedly yells 'You get $200 if you catch two cats!' from the back of the room, and everyone starts laughing. Hunting cats is clearly a source of pride for this team.

Often, the rangers find hopping mice and small birds in the cat guts, but once they found a nyinytjirri (rough-tailed lizard), which has a very spiky tail. Studying the gut contents can help rangers target their management and hunting efforts. In just a few months at the start of this year, the community caught 15 cats. '[They're] gonna send us broke!' Dannica laughs.

Hunting cats in the Kiwirrkurra style takes a lot of time and energy, so new methods are being utilised to tackle the problem on a wider scale. To kick off new management strategies and research initiatives, the West Australian government committed $7.6 million as a part of the WA Feral Cat Strategy, including $500 000 in grants. The Kiwirrkurra IPA rangers received some of this grant money to undertake a project to protect a remote ninu population.

Using Eradicat – a sausage laced with 1080 poison – the rangers are going to see how this strategy impacts cat populations and whether it increases ninu survival. This area was selected by Traditional Owners and the ranger team because it has no permanent water source, meaning there are very few dingoes around – it is just cats and foxes. It's an exciting mission for the rangers as they get to share their knowledge and save the ninu, something that they're very proud of.

Caring for ninu and tjalapa is a particularly important aspect of tjukurrpa, and the Kiwirrkurra IPA ranger team are working tirelessly to achieve this. Recently, a lot of ninu have been seen around the Kiwirrkurra community. There's some guesses that this could be due to the strong start to cat hunting efforts at the start of the year.

'Bilby [are] coming up everywhere,' says Yalti.

❋ *Smooth fade*, p. **188**
Skink on the brink, p. **248**

WHEN DID YOUR KIDNEY STONE START GROWING? ANSTO SCIENTIST CARBON DATED HIS TO FIND OUT

Jacinta Bowler

In 2011, Vladimir Levchenko experienced the searing pain about six per cent of us will experience – a kidney stone beginning its excruciating trek down the urinary tract.

A trip in an ambulance to the hospital led to him being wheeled into an operating theatre to have his kidney stone removed.

In between the bouts of pain, Dr Levchenko, who is a research scientist at Australian Nuclear Science and Technology Organisation (ANSTO), began wondering when his problem began.

He was curious to see if he could work out how long ago his stone had started forming. And as an expert in carbon dating, he began asking questions about his stone.

When the doctor said it was made of a carbon compound called oxalate, 'that clicked in my brain. I can [date] that', Dr Levchenko recalls.

He'd recently been working on dating oxalates from Aboriginal rock art.

'I asked, "Could you please save it for me?" And he did.'

This fateful moment led to multiple papers, and a deeper understanding of how kidney stones grow. But Dr Levchenko thinks there's plenty left to do.

How kidney stones grow

Kidney stones are formed from calcium salts produced by our urine. While most of this is removed from the body when we go to the toilet, occasionally the salt can become a tiny crystal that grows over time.

Urology surgeon and kidney stone expert Gregory Jack's first interaction with a patient is usually when they show up at the emergency room with a kidney stone in their urinary tract.

'You've got something too big being pushed through a tube that's too small. Almost like a golf ball through a garden hose,' he says. 'They've been unaware of it [before the pain started], and they're quite traumatised by the pain. And I won't have any old scans or any old records.'

Once someone has been diagnosed with kidney stones they are normally tracked over years or decades using X-ray or ultrasound, but finding the first stones can be a surprise for both the patient and the doctor.

If the stones are small, patients will pass them through their urinary tract, but larger stones are removed via surgery, or broken up with ultrasound. Once the stone has exited the body, it's possible to track how old it is – approximately – by understanding how it formed and looking at its 'growth rings' as more layers of calcium salts are added.

'Over years, decades or even a person's lifetime, the stone will slowly grow,' Dr Jack says. Similar to a tree, 'we crack open the stone, and often we'll see 100 different rings inside'.

Fast kidney stones with extremely thick rings can grow in a few months, while most take years or decades.

'Some years the rings are quite fat, meaning the stone grew quite quickly. Other years, the rings may be quite thin, meaning the patient was having a good year.'

But they're not quite as accurate to years as tree rings, meaning they don't provide an exact date, and don't give doctors or patients information on how the stones may have started. This means doctors aren't able to provide information to patients about how long it could take for another stone to form and pass.

Carbon dating using the bomb pulse

Once Dr Levchenko had recovered from surgery, he took his 6-millimetre kidney stone, and removed a tiny sample of the stone's core, middle and outside.

'The biggest difficulty was just to find out where it started growing. It's not necessarily symmetrical,' he says.

These tiny samples were then analysed by a giant machine called an accelerator mass spectrometer. This process measures the amount of a radioactive isotope called carbon-14 inside the kidney stone samples to determine their age. The technique is also used to date ancient bones, plants or anything else that's absorbed carbon over its lifetime.

When the organism dies, fresh carbon stops being exchanged with the environment, and the carbon-14 begins to steadily decay, allowing scientists to estimate the time of death.

And while specimens under 60 000 years old can be dated down to a few hundred or thousand year time frame, specimens after 1950 – such as Dr Levchenko's kidney stones – can be dated much more accurately. That's because of something called the 'bomb pulse' – when carbon-14 in the atmosphere doubled as a result of nuclear bomb tests in the 1950s and 1960s. This increase in carbon-14 allows researchers to date samples precisely, sometimes all the way down to the month.

Being able to use this bomb pulse to accurately date samples will only last until about 2030, when carbon-14 levels return to baseline.

This pulse allowed Dr Levchenko to date the kidney stone extremely accurately from the outer sample to the core.

The outer edge sample correlates to when the stone was removed. Then working backwards to the core, he discovered his stone had been slowly growing inside his kidney for 17.6 years.

This finding led him to undertake even more research with a group in the Netherlands, with two more stones sent over from two separate patients, which also provided some interesting results. Those two stones, which were similar in size, had completely different growth rates – one had grown over the past 23 years, while the other had grown in just seven years.

'Knowing the growth rate is helpful, because it does help counsel the patient. If it took 15 years to grow one, you can extrapolate 15 years for the second,' Dr Jack says. But, he adds, 'knowing the timing is fun, but it probably doesn't help the prevention as much as knowing the cause'.

How to prevent kidney stones

When a kidney stone is passed from the body, normally a number of tests are done to test the stone, to try to work out what caused it.

'The first thing we'll do is we'll send it to the lab to see what type of stone it is,' Dr Jack says. 'The vast majority will be a calcium stone, but there'll be other types ... Knowing the type will help us determine a little bit of what's going on.'

Unfortunately, Dr Jack has seen plenty of research trends come and go when it comes to how to prevent kidney stones.

'Our poor patients. A decade ago they were told to not have calcium. And then we realised we got that all wrong. So then they were told, don't have oxalates [found in some foods such as leafy green vegetables]. And then we found out we got that wrong too,' he says. 'The diet stuff just comes in fads, and then all of a sudden, everybody cuts out spinach and kale.'

Now there are three things Dr Jack advises almost all his patients:

- drink more water
- have less salt
- eat less animal protein.

The rest of the advice is normally tailored to the individual.

'We can do fancy urine tests and measure out in someone's urine a number of different electrolytes and see what levels are off in that individual,' he says. 'I see all of these patients every day, and they all want to know what caused [their stone].'

'Whatever comes naturally'

Dr Levchenko has since had one more, smaller kidney stone, but this one was seen on an X-ray screen, and removed before it became a problem.

But the original kidney stone dating, and subsequent publication of his research, brought him in contact with kidney stone researchers around the world.

Despite the research he's already done, Dr Levchenko believes that there's plenty more to be done in this space. According to him, further research may help scientists understand more about how diet and lifestyle affect stone growth.

While he had plans to work with the Dutch researchers to investigate more stones, COVID caused the plans to be shelved.

'When the pandemic subsided, the research structure, organisation and funding changed too,' Dr Levchenko says. 'In Australia, there's no active groups in this field ... Hopefully, it will change.'

He says that when he first started getting in touch with kidney stone researchers, 'they didn't understand what I was suggesting'.

They assumed he wanted to feed their patients radioactive isotopes.

'I said, "No, they eat them anyway. It comes in our diet",' he laughs. 'I'll measure whatever comes naturally.'

✱ *Faster higher stronger doper*, p. **74**
   Mirror molecules: The twisted problem of chemical chirality,
        p. **104**

CHINA'S RACE FOR FUSION ENERGY

Gemma Conroy

On a cold February morning in Hefei, the snow-blanketed grounds of the Chinese Academy of Science's Institute of Plasma Physics (ASIPP) are unusually quiet. China's New Year is approaching, and most people in the city are preparing for days of dragon-themed celebrations. But inside the institute, researchers are still hard at work. In a vast control room under a ceiling studded with red neon-lit stars, plasma physicist Xianzu Gong is taming a different kind of fiery beast.

Gong's dragon is a fusion research reactor: the Experimental Advanced Superconducting Tokamak (EAST). Tokamaks are doughnut-shaped machines that generate the same nuclear reactions that power the stars. They use magnetic fields to confine heated loops of plasma – a fluid-like state of matter containing ions and electrons – at temperatures hotter than the Sun's core. The aim is to force atomic nuclei to fuse, releasing energy. This could be harnessed as a source of almost limitless clean power, if the scorching, unstable plasma can be maintained and controlled for long enough – a feat yet to be accomplished.

Corralling the unruly plasma is gruelling work. Every day, Gong and his colleagues fire up around 100 shots of plasma from early morning until around midnight. By comparison, the Joint European Torus (JET) in Culham, UK, which was the world's largest fusion-research facility before it closed last year, achieved 20–30 shots each day. 'Almost no weekends, no holidays for us,' says Gong, who heads EAST's physics and experimental operations.

Although only a stepping stone to anticipated fusion power plants, EAST is one of the facilities that's putting China on the map in the global race for nuclear fusion.

The world's most well-known fusion experiment is the US$22 billion International Thermonuclear Experimental Reactor (ITER), a giant tokamak being constructed in southern France, to which China is contributing. And in recent years, ambitious firms in the United States and elsewhere have raised billions of dollars to build their own reactors, which they say will demonstrate practical fusion power before state-led programs do.

At the same time, China is fast pouring resources into its fusion efforts. The Chinese government's current five-year plan makes comprehensive research facilities for crucial fusion projects a major priority for the country's national science and technology infrastructure. As a rough estimate, China could now be spending $1.5 billion each year on fusion – almost double what the US government allocated this year for this research, says Jean Paul Allain, associate director of the US Department of Energy's Office of Fusion Energy Sciences in Washington, DC. 'Even more important than the total value is the speed at which they're doing it,' says Allain.

'China has built itself up from being a non-player 25 years ago to having world-class capabilities,' says Dennis Whyte, a nuclear scientist at the Massachusetts Institute of Technology (MIT) in Cambridge.

Although no one yet knows whether fusion power plants are possible, Chinese scientists have ambitious timelines. In the 2030s, before ITER will have begun its main experiments, the country aims to build the China Fusion Engineering Test Reactor (CFETR), with the goal of producing up to 1 gigawatt of fusion power. If China's plans work out, a prototype fusion power plant could follow in the next few decades, according to a 2022 road map published in *The Innovation* ('Recent progress in Chinese fusion research based on superconducting tokamak configuration').

'China is taking a strategic approach to invest in and develop its fusion energy program, with a view of long-term leadership in the global field,' says Yasmin Andrew, a plasma physicist at Imperial College London.

Building artificial suns

Scientists have been trying to make fusion reactors work since the 1950s. The idea is to merge two hydrogen nuclei – which are positively charged and therefore repel each other – into a larger helium one. In the Sun, gravity generates enough pressure to do this; on Earth, high temperatures and strong magnetic fields are necessary. So far, however, researchers haven't been able to keep fusion reactions running long enough to produce more energy than it takes to spark them.

In late 2022, researchers at the US National Ignition Facility (NIF) in Livermore, California, announced a breakthrough when they briefly recovered more fusion energy than they put into their target. Using an alternative design to a tokamak, NIF fired 192 laser beams at a tiny pellet of the hydrogen isotopes deuterium and tritium, causing them to fuse. However, much more energy went into operating the lasers than was delivered to the target. Many researchers say the most practical approach to fusion energy will entail using a tokamak to confine a long-lived 'burning plasma', one in which the fusion reactions provide the heat needed to sustain it. One of ITER's targets, seen as a general prerequisite for viable fusion plants, is to create a burning plasma that produces ten times the power that went into it.

If scientists can do this, fusion could offer a safer, cleaner alternative to conventional nuclear fission power plants that split heavy uranium nuclei, producing radioactive waste that can remain dangerous for thousands of years. Fusion reactors would produce only short-lived waste. Another safety feature is that fusion reactions simply stop if the plasma falls below a certain temperature or density. And the process is expected to be more efficient than fission; the International Atomic Energy Agency says that fusion could generate four times more energy than does fission, per kilogram of fuel.

It's a particularly tantalising prospect for China where, between 2020 and 2022, several regions experienced massive power outages owing to skyrocketing demand for electricity during frigid winters. Despite rapid progress in renewable energy, the country still generates more than half of its electricity from coal and remains the biggest contributor to global carbon emissions. And although China is aiming to achieve peak emissions by 2030 and carbon neutrality by 2060, its energy requirements are set to double over the next three decades. 'We need innovations that reduce carbon – that's our dream. Nuclear fusion energy can do this,' says plasma physicist Yuntao Song, ASIPP's director-general.

China's vision

In EAST's control room, Gong prepares to fire another pulse of plasma with a click of his mouse. The plasma itself lies behind the control room's wall of monitors, confined in a vacuum chamber that has the Chinese flag mounted on its roof. 'Every shot could be in support for the future of fusion energy,' Gong says.

China's involvement in fusion began with building several small and medium-sized tokamaks using components from devices in Russia and Germany. In 2003, it joined the international ITER experiment, alongside the European Union, India, Japan, Korea, Russia and the United States.

In 2006, China opened EAST, which has since racked up world records for sustaining plasma lasting minutes, instead of seconds. EAST's knack for creating long-lived plasmas has made it an experimental workhorse for ITER, particularly for quickly cross-checking results, says Alberto Loarte, who heads ITER's science division. 'The research in China is extremely dynamic,' he says.

Loarte cites how, in January 2024, he and his colleagues spent a week running experiments at EAST, to verify that lining a reactor's plasma-facing walls with tungsten can achieve a tightly confined plasma, even if the walls aren't also coated with a boron layer to keep out impurities. (These findings will help ITER, at which in October 2023 researchers decided to switch wall linings to tungsten instead of beryllium.) In many countries, such an effort would have taken months to organise, says Loarte. But in China, plans often come together in weeks because many research groups don't require formal proposals or lengthy discussions to get to work.

ITER originally aimed to start experiments in 2020 but has been plagued by delays. In July 2024, researchers announced that it will push back its major experiments to 2039. Most ITER countries are developing their domestic fusion capabilities in parallel, but few are doing so as intensively as China, says Jerónimo García Olaya, a fusion scientist at the French Alternative Energies and Atomic Energy Commission in Paris. 'They are building a very ambitious program,' says Olaya, who co-leads experiments at JT-60SA in Naka, Japan, currently the world's largest tokamak in operation.

Among China's other research fusion reactors, besides EAST, is its HL-3 tokamak, opened in 2020 at the Southwestern Institute of Physics in Chengdu. Experiments at China's facilities will feed into the next-generation CFETR, although construction still needs approval from the government. An official at ASIPP who didn't want to be named couldn't give a timeline for this, but says that the government is factoring ITER's timeline into its decision. The CFETR, which will be slightly bigger than ITER, aims to bridge the gap between ITER – a purely experimental device – and demonstration plants that would generate electricity.

CFETR first aims to generate between 100 and 200 megawatts of net power, producing more power than went into heating the plasma, but not enough to cover the electricity used to operate the facility. By the 2040s, its goal is to deliver more than ten times as much heat as is directly put into the plasma, the milestone for viable fusion, and also to produce up to a gigawatt of net power. If this could be achieved, demonstration power plants would then produce grid electricity.

CFETR's engineering design report, released in 2022, places the facility ahead of several demonstration power plants, including the European Union's and Japan's proposed DEMO reactors – expected to begin their engineering designs in 2029 and 2025, respectively.

China's strength in fusion research lies not so much in stand-out engineering innovations, says Allain, as in its speed and focus on developing the materials, components and diagnostics systems needed to build reactors.

To develop CFETR, ASIPP has started building a sprawling 40-hectare workshop (about the size of 60 football fields) a short drive from EAST. Scheduled for completion next year, the Comprehensive Research Facility for Fusion Technology (CRAFT) is a massive hub where researchers will develop and manufacture materials, components and prototypes for CFETR and subsequent fusion power plants.

In the United States, a similar facility to develop key fusion technologies has been flagged as a priority for years, but plans have failed to materialise, owing to limited funding and other issues, says Whyte. 'It has been frustrating,' he says. 'There are positive signs of change, but we lost our lead.'

China's focus on building a fusion workforce has also given the country an edge in personnel, says Hongjuan Sun, a plasma physicist at the UK Atomic Energy Authority in Abingdon. 'They really put a lot of effort in training the next generation,' says Sun, who worked on JET. Allain estimates that China has thousands of PhD students in fusion, compared with mere hundreds in the United States.

Commercial efforts

Although China's program is ramping up fast, start-up firms around the world make much bolder claims about the pace with which they can commercialise fusion energy.

For example, Commonwealth Fusion Systems (CFS), a spin-off from MIT, promises that its tokamak, called SPARC, will be the first to churn out more fusion energy than the heat that the plasma consumes. The firm, which is based in Devens, Massachusetts, and is working with MIT researchers, says SPARC will produce its first plasma by the end of 2026. The effort relies on advances in high-temperature superconducting materials, which should allow the tokamak to be much smaller and quicker to build than ITER and other giant facilities. CFS says it will have plants supplying electricity grids by the early 2030s. Other firms are making similarly bullish statements about various designs for pilot fusion plants.

Globally, more than 40 companies are working to commercialise fusion, and together have received investments of $7.1 billion, says the US-based Fusion Industry Association (FIA).

But China's industrial efforts are burgeoning, too. The country's fusion start-ups have attracted more than $500 million in investment in just a few years, says Andrew Holland, chief executive of the FIA. That places China second only to the United States, which has poured more than $5 billion into fusion companies. 'The private fusion effort in China is significant,' he says.

In January 2024, the Chinese government launched a national consortium called China Fusion Energy. Led by the China National Nuclear Corporation, it brings together 25 government-owned companies, four universities and a private firm with the goal of pooling resources to accelerate China's fusion effort.

Among industrial heavyweights in fusion research is the ENN Group, one of China's biggest private energy conglomerates. According to the FIA, the company has invested more than

$200 million in its fusion energy program. An ENN road map envisages building a 'commercial demonstration' reactor by 2035.

A handful of dedicated fusion companies have sprouted up in China over the past three years. Among them is Energy Singularity, a Shanghai-based start-up founded in 2021 and the country's first dedicated fusion power firm. Much like SPARC, Energy Singularity aims to build smaller, less expensive tokamaks by taking advantage of the latest materials for magnets; it has so far attracted around $110 million in funding, says co-founder Zhao Yang. In June 2024, the firm's HH70 tokamak achieved its first plasma using high-temperature superconducting magnets – a world first, Yang says.

Energy Singularity is planning a next-generation device, HH170, which aims to produce ten times more energy than the heat needed to fuel the plasma. Just as optimistically as the US firms, Yang estimates that the small tokamak will take only three to four years to build, instead of decades.

One of the big questions in fusion surrounds the availability of fuel. For tokamaks, a mixture of deuterium and tritium (D–T) isotopes is considered one of the most efficient fuels. But tritium occurs in minuscule traces in nature, so will need to be produced in fusion facilities, through a reaction between the neutrons produced during fusion reactions and a blanket of lithium in the tokamak wall. Whether such 'tritium breeding' can actually work is unclear.

ITER is one of the largest research efforts that will explore this question. But China has speedier plans: its Burning Plasma Experimental Superconducting Tokamak (BEST), built next to CRAFT and due to be completed in 2027, will also run D–T experiments and explore whether tritium can be bred, says ASIPP director Song.

It's all part of a long-term push to develop what many see as a key solution to the world's energy problems. Back at EAST, in contrast to the bullish claims of private firms, Gong sees the race for fusion energy more as a marathon than a sprint. He has thousands of plasma shots ahead of him. 'There's still a lot of work we need to do,' he says.

✳   *Sounds of the slow-rolling sea*, p. **175**
    Ethical problems continue to plague biometric studies of
        Chinese minority groups, p. **215**

'EARTH POETRY' IN THE ARCTIC

Lucinda Duxbury

My favourite place isn't a fixed location in space.

My favourite place is aboard a research vessel that sails the seven seas to drill down into the soft sediments lying at the bottom of the ocean. Contained in the mud are traces of past climates and environments, laid down over millions of years.

When we analyse these traces, they offer a portal into Earth's past. And if we pinpoint particular time periods, they're a muddy window into a warmer future.

Into the floating lab

I'm in the far corner of the microbiology lab, dressed head to toe in a white suit, mask and safety glasses. All you can see are my eyes.

Right now, I'm focused on the task at hand. I'm cutting off the tips of hundreds of sterile plastic syringes so we can plunge them into the kilometres of fresh sediment cores we unearth from the seafloor.

Cut, clean, bag, seal, rinse, repeat. It's monotonous but I'm leaning into it. Repeat, repeat. Repeat like the past repeats. It feels like a form of meditation.

Window to another world

I leave the lab and look through the window to the world outside. Svalbard's snow-capped fjords to one side of us, sea ice to the other.

We're well past the Arctic Circle now. Into the fog and into the thick of it. A sea of ice, whales, porpoises and puffins.

The concept of time is slippery under an unrelenting midnight sun. If you think about it, the Arctic summer is just one big, long day (with lots of naps in between).

I imagine it should be peaceful here. But we come in a loud diesel-powered boat bearing news of an ice-cracking climate crisis.

For two months, a shipload of scientific minds whirr around the clock. Geological time blurs as we haul up rock dropped millions of years ago from the melting icebergs of ancient ice sheet retreat.

I've come all the way from Tasmania, an island falling off the bottom of the world.

I've never been this far north. But in this closed-off clean corner of the ship, I sometimes forget it.

The flow of heat

We're about to arrive at our first drill site. Our journey on this expedition will track the path of the West Spitsbergen Current into the Arctic Ocean. The current acts as the main way heat from the warmer lower latitudes is transported to the deep north.

We're collecting sediment cores that will help us understand how this heat source has interacted with Arctic ice, atmosphere and ocean through time. What happens with heat flow into the Arctic affects the global system.

For example, increases in meltwater can slow overturning ocean currents, with dramatic effects for northern hemisphere climate. If we study this system in the past – especially during times of elevated temperatures and CO_2 – we might be able to glimpse into our future.

Traces of DNA

The excitement as we start drilling on the first site is palpable. An international goliath effort has gone into just getting us to the point where we can begin.

It's like Christmas Eve. I can't sleep so I write little poems to pass the time. We play the epic and dramatic song 'Ecstasy of Gold' on the loudspeakers to welcome our first sediment core on deck. I'm ready in my white suit, gloves, mask and glasses – sterile syringes in hand.

There are secrets in the sediments. I'm looking for ancient DNA in my samples, potentially millions of years old, to reconstruct traces of past ecosystems.

The DNA is so degraded and broken down, we must go to extreme lengths to protect it from contamination from modern, intact DNA that would drown out the ancient signal. That's why I'm wearing all this protective gear. We don't have to protect ourselves from the samples, we must protect the samples from us.

Back on land we will sequence this DNA, generating gigabytes and gigabytes of genetic code. We will compare our data to a reference database of known genetic sequences to work out what phytoplankton and zooplankton made up the marine ecosystem when the Arctic was ice-free in the past.

Marine microorganisms such as plankton form the base of entire food chains. They affect the cycling of greenhouse gases including carbon dioxide through the Earth system. Understanding the composition of these past communities has huge implications for Earth's future.

Earth poetry itself

After a few weeks we settle into the routine of life on the ship. I go to bed knowing I may be woken up at any hour if we start drilling a new hole.

For two mad months we give ourselves permission to be totally absorbed by our science. With this comes a communal depth of knowledge made possible only by circumstance and this ship.

Sometimes, though, the glitz and glimmer wear off and our patience with each other wears thin. What was once shiny shows its true colours – and it's the uninspiring hue of green–brown mud.

Through these moments I cling to the words of the great marine biologist Rachel Carson: 'The sediments are a sort of epic poem of the earth.' Downwards we dive into the tremulous echoes of the past.

Sometimes when I'm in my cabin trying to sleep, the low grumbles and creaks of the ship could almost be mistaken for the heave, groan and grating of the ancient ice sheet herself – to the beat of the climate cycles. Her echoes we bring to life with X-rays and smear slides.

Until the end

We're being chased south by ice now. Our ship is not an icebreaker so whatever move she makes, we acquiesce.

Under the harsh light of an Arctic summer, her twilight draws closer. The Intergovernmental Panel on Climate Change (IPCC) predicts with high confidence that we will experience at least one ice-free Arctic summer by 2050.

Then, one day, the expedition is over.

Where we once watched a live cam of the drill floor, we stream the skateboarding at the Olympics.

Soon enough we're back in busy shipping lanes. Back in shorts.

Night, too, returns. Out on deck a group gathers to watch for the first time, in a very long time, as the sun slips unceremoniously below the horizon. It signals a welcome restoration of balance in our lives.

This was the final expedition for the RV *JOIDES Resolution*. After decades of service to the science community, the ship herself is an archive of story and adventure. As you read this, her labs have already been dismantled. But she won't be fully gone for a long time.

The *JR* will have a half-life in our memories of far-off places and times, in the moments that become enshrined in legend as they are retold and re-remembered, at times misremembered. Stories from the people that worked and lived together – beautifully and messily – to uncover oceanic secrets.

✳ *Whale talk*, p. **159**
   Skink on the brink, p. **248**

FISHING FOR A GLACIER'S SECRETS

Mark Horstman

It's an ice fishing trip like no other.

A 5-metre long drill. More than a kilometre of line. On a glacier tongue over an unmapped ocean.

In the summer of 2025, glaciologists Sarah Thompson and Maria 'Coti' Manassero from the University of Tasmania are in East Antarctica to plumb the depths beneath the floating ice shelf of the Scott Glacier. Their mission: to suspend a string of sensors under the ice that will monitor the temperature, salinity and currents of the ocean below, every hour for the next few years.

Around 5000 kilometres southwest of Australia is a vast 'verandah' of interlocking ice shelves and glacier tongues, extending over the ocean and covering more than half the area of Tasmania. One of the largest in East Antarctica, it's the most northerly ice shelf system outside of the Antarctic Peninsula.

The Shackleton ice shelf buttresses glaciers like the Scott and Denman, restraining their flow of ice into the sea. But for glaciers like the Denman, their 'grounding lines' (where the ice sits on bedrock) appear to be retreating each year as warm ocean waters melt them from below.

In most places the ice is many hundreds, even thousands, of metres thick. So to get to the ocean underneath to measure its properties, you need to find a 'window' that opens into the water body you want to investigate.

Enter Duanne White, a geomorphologist from the University of Canberra.

Satellite imagery and field reconnaissance led him to discover an 'epishelf lake' – a frozen freshwater lake formed when meltwater flowing off a glacier is trapped behind a floating ice shelf.

'It's a small window of sea ice right near the grounding line of the Scott Glacier, tucked in behind rocky Cape Hoadley,' he says. 'It's the ideal site because it gives access to the southernmost part of the ocean cavity beneath the ice shelf without having to drill through more than 600 metres of glacial ice, so you can use tools designed for ice fishing on lakes in North America.'

The trick is to install a mooring line at the right spot via a hole drilled through the ice.

'Here you can measure the speed and characteristics of meltwater flowing out from the base of the ice shelf, right up near the grounding line, where ice from the Scott Glacier lifts off the bed and starts to form the ice shelf,' says Duanne.

In East Antarctica, the main changes to the ice happen at the interface between the ice shelf and the ocean, where melt rates are thought to be highest.

Are warm waters reaching right to the back of the ice shelf? That's what the fishing team plans to find out.

Sarah, Coti and polar field guide Nick Morgan from the Australian Antarctic Division camped on the frozen lake, where they knew the ice was thick enough to work on but easier to drill through.

'It was a really beautiful site to stay as there were pressure ridges from the lake ice on one side of us and cliffs on Cape Hoadley on the other side. We were also very lucky with the weather and, once the morning katabatic winds dropped, had great conditions to work in,' says Sarah.

Using a battery-powered auger, over a few hours they drilled a hole about 25 centimetres wide.

'Luckily, the ice was 4.8 metres thick, as we only had five metres of drill flights!' Sarah says. 'We drilled one hole, and once we were through, we sent the camera down to get an estimate of depth and to see what the bottom looked like.'

Coti says it was also fortunate they brought 1500 metres of cable.

'We were astonished to find depths exceeding 1200 metres below sea level, with rocky and rugged slopes,' Coti says. 'We were also amazed by the abundance of marine life we saw with the camera, like sponges and brittle-stars.'

This is where the hard work came in: reeling the camera and sediment core up and down for over 1200 metres, taking about three hours each time.

The only thing that didn't go to plan was the sediment core. They expected to retrieve some sediments, but couldn't collect any due to the rocky slopes on the seafloor.

Sarah agrees the depth of the ocean was surprising, given they were so close to the grounding line. 'We expected it to be quite deep, but not close to 1300 metres! The site is only a few hundred metres away from the shore, so it must be a really steep drop-off.'

The sensors on the mooring line were carefully lowered to a depth of 690 metres and secured to a frame with a solar power source and transmitter at the surface.

'The hardest part for me was being confident that I'd done it all correctly!' laughs Sarah. 'I checked everything three times, but we didn't have confirmation until we got back to the main field camp with internet and started to receive data. That was a huge relief!'

Back in Hobart, oceanographer Madi Rosevear at the University of Tasmania is thrilled to be in touch with the mooring station.

'Every day at 11 am I'm getting four emails from Antarctica with hourly measurements of ocean temperature, salinity and current speed. Hopefully, this will continue for the next two to three years,' Madi says. 'These data will tell us how much heat is available to melt the nearby Scott Glacier ice tongue and will give important insights into the ocean circulation, including how it changes over time.

'The discovery of seafloor more than 1200 metres deep at this location, close to the Scott Glacier, is very significant. Deep troughs and canyons around Antarctica provide pathways for warm, salty Circumpolar Deep Water to access ice shelves, where it can drive rapid melting.'

It ties in with a single data point showing Circumpolar Deep Water under Shackleton ice shelf collected in fieldwork last year.

For Sarah the successful installation means invaluable science infrastructure in a location not reached before.

'If warmer ocean water can reach the point at which the glacier starts to float, there is significant potential to accelerate melting and speed up the glacier flow, transferring more ice into the ocean. By installing the mooring, we can not only measure what is happening at this site, but we can monitor any changes in the next few years and start to predict the stability of the system with greater confidence.'

Even better, says Madi, mooring data from the terrestrial side will dovetail with upcoming research from the marine side. 'Having the mooring in the water making measurements during the Denman Marine Voyage this year is extremely powerful, since we'll be able to compare the water properties inside and outside of the ice shelf cavity.

'Hopefully, we'll be able to assess the circulation pathways beneath the ice, and ascertain how vulnerable the ice shelf is to changes in the ocean around Antarctica.'

Sarah says it will provide the vital link between what we see just offshore in the ocean and how that might be affecting the ice shelves. 'Because we don't have a lot of information about the shape and depth of the ocean floor beneath the ice shelves, we don't know whether the warmer water masses we see offshore can reach the grounding line and contribute to melting at the base of the ice.

'With the mooring in place close to the grounding line of Scott Glacier we can directly compare, in real time, the ocean properties there with the offshore measurements profiles that will be collected as part of the marine voyage on Australia's icebreaker RSV *Nuyina*.'

IS THIS ACTUALLY PTSD? CLINICIANS DIVIDED OVER REDEFINING BORDERLINE PERSONALITY DISORDER

Natasha May

When Professor Andrew Chanen was a trainee psychiatrist in 1993, patients with borderline personality disorder (BPD) who had self-harmed were 'vilified' and 'treated appallingly'.

'There was this myth that somehow they were indestructible,' he says. Despite what his teachers told him, 'most were dead by the end of my training'.

More than three decades later, Chanen is the chief of clinical practice and head of personality disorder research at Orygen, the National Centre of Excellence in Youth Mental Health at the University of Melbourne, and he says BPD remains the most stigmatised and discriminated against mental health disorder in Australia and internationally.

Overwhelmingly diagnosed in women, BPD is characterised by difficulty managing emotions, rapid mood changes, self-harm often accompanied by suicidal thoughts, and an unstable self-image.

Some Australian clinicians are calling for BPD to be recognised as a trauma disorder rather than a personality disorder, arguing this would lead to better treatment and outcomes.

The argument for rethinking BPD

American psychoanalyst Adolph Stern introduced the word 'borderline' to psychiatric terminology in 1938, using it to describe a group of patients who fitted neither the neurotic nor the psychotic diagnostic categories.

Several studies have shown BPD is associated with child abuse and neglect more than any other personality disorders, but the rates can vary from as high as 90 per cent to as low as 30 per cent. An analysis of 97 studies found 71.1 per cent of people who were diagnosed with the condition reported at least one traumatic childhood experience.

Dr Karen Williams, who runs New South Wales's Ramsay Clinic Thirroul – Australia's first women-only trauma hospital – believes BPD 'is a gendered diagnosis that is given to women who have got histories of abuse, whereas when we see a man come back from a traumatic event, we [say] he's got PTSD [post-traumatic stress disorder]. There is no symptom that a borderline personality disordered person has that a PTSD patient doesn't also have.'

Williams says it often takes several sessions before she can uncover a patient's abuse. The response of dissociation and forgetting trauma is very common, she says. Also, not all patients recognise their experiences as trauma.

Despite there being no clinical difference between PTSD and BPD, Williams says the clinical response varies markedly. PTSD, particularly among veterans, is treated with sympathy, while women with the diagnosis of BPD are considered 'difficult'.

Williams prefers the term 'complex post-traumatic stress disorder' to BPD, as does Professor Jayashri Kulkarni, the director of the Monash Alfred Psychiatry Research Centre. Kulkarni says the BPD label implies the behaviour is part of a personality style. There's an implied 'stern moralistic approach' that these people should just be able to control themselves – and that attitude contributes to stigma.

But she says the more she has researched BPD, 'the more obvious it seems the women and the men who have been labelled with this condition often have dreadful early life trauma. I really think this is injustice, to say to somebody who's gone through hell in their early life and onwards, that they've got a significant flaw of their inner core.'

The case for the term 'personality disorder'

To Chanen, the term 'personality disorder' is useful because it captures the identity and relationship difficulties he says are at the heart of the issue.

He points to a national study of childhood maltreatment published in 2023 which showed nearly two-thirds of the population experience some form of childhood adversity. Despite that, BPD is comparatively rare, occurring in only 1 to 3 per cent of the population.

'There's something important going on in each individual that interacts with the experience of adversity. While that interaction might give rise to borderline personality disorder, it might also give rise to another disorder, such as depression, or no mental disorder,' he says. 'That's not to say that the adversity is unimportant, but it's not inevitable that a person will develop a mental disorder, and certainly not inevitable that they will develop borderline personality disorder.'

Chanen believes any reductionist arguments about causes are 'oversimplified, wrong and unfortunately harmful for people living with personality disorder'. He believes the debate around renaming the disorder as complex PTSD is 'not really supported by the science and weakens the moral argument for respect, dignity and equality of access to effective services'.

Chanen is concerned a name change may have the unintended consequence of invalidating the experiences of patients who have not experienced trauma, or prompt clinicians to assume that trauma is present without any evidence. Instead, he believes early intervention is key.

An associate professor at the University of Sydney, Loyola McLean, who identifies as a Yamatji woman, says of the divided opinions within her profession: 'It could well be that we're talking about two halves of the same whole.

'I think we've got to keep an open mind that this adverse experience may be contributing, triggering, and for some people will have a causal element,' says McLean, who is a consultation-liaison psychiatrist and psychotherapist.

'Trauma – in particular early trauma, because that's where the body and brain are really developing – we know that it's such a huge risk factor for downstream health problems across the spectrum of health problems.'

The physical and the psychological are deeply connected, she says, but 'the whole of the western world is still suffering from a kind of a Cartesian divide'.

A shifting approach

The discussion about using the term 'BPD' or 'complex post-traumatic stress disorder' is about more than words – according to Kulkarni, it changes the whole direction and focus for treatment.

Historically, psychiatric treatment for BPD has relied upon antidepressants to treat low mood and antipsychotics for paranoid thinking, but it has not addressed underlying cognitive symptoms such as difficulty managing emotions, a disturbed sense of identity, disturbed relationships and impulsivity. Those symptoms tend to be treated with psychosocial approaches, such as dialectical behaviour therapy, mentalisation-based treatment and high-quality care.

Kulkarni and Dr Eveline Mu at Monash Alfred Psychiatry Research Centre are running clinical trials for new drugs to target the neurochemistry they believe drives the symptoms of BPD/ complex post-traumatic stress disorder.

The effects of trauma on the body's stress levels mean the glutamate system – the primary neurotransmitters of the nervous system – is in overdrive, Mu says. Her theory is that this drives cognitive dysfunction.

Since it began in 2022, 200 people have participated in the randomised controlled double-blinded clinical trial of memantine, a drug that the regulator has approved for treatment of Alzheimer's patients, and which blocks the body's glutamate receptors.

Williams' women's-only trauma hospital is also examining new ways of responding to those with acute symptoms. She says the only place where acutely suicidal patients can go are mixed-gender rooms in hospital psychiatric wards, which have no locks and can lack supervision of male patients who are often psychotic, drunk and detoxing. Sexual assault is often rife in such wards.

It's an environment that exacerbates symptoms, she says.

By contrast, the three-week program her patients undergo involves exercise, self-care, and education about healthy relationships.

'Almost all the time, they don't just have trauma from their childhood, but they've still got it now,' Williams says. 'We know that people who have been abused tend to end up in abusive relationships again, because they have such little self-value and they don't know that they deserve to be treated better.'

The hospital's beds are constantly full with patients who can afford private treatment, with some even coming from interstate. Only one of the hospital's 40 beds is publicly funded.

Williams says her program has improved the quality of life of her patients, with many able to take on full-time work or go back to study. 'Many of them have said: "I want to be a nurse, I want to come back and work here."'

Kulkarni says one of the other new solutions is to get rid of the label. 'It's hurting people ... Taking a new look offers us new compassion and new understanding.'

Author's note: Since writing this piece, Karen Williams has left Ramsay Clinic Thirroul, after raising concerns that the clinic was no longer completely trauma focused and also treated people with substance addiction alongside them. I discussed this in a follow-up article, 'NSW's sole women's-only clinic dedicated to trauma ends "gamechanging" focus' in the *Guardian* in April 2025.

✻ *Some psychedelic medicine developers want to ditch the therapy*
aspect. What could go wrong?, p. **120**
Why can't we remember our lives as babies or toddlers?, p. **204**

A FREEDIVER FINDS BELONGING WITHOUT BREATH

Sally Montgomery

Face down in deep water, I float. I inhale through a snorkel. My eyes are closed. Body relaxed. I raise one finger, motioning to my dive buddy that I am taking my last breath. Finally, after several minutes of floating meditation, I am ready.

I exhale all the air I possibly can.

My chest deflates. An uncomfortable tightness grips.

Grips me.

I take a single breath.

It is deep and full.

My chest expands. My stomach too.

I feel it in my throat.

With a splash, I duck-dive, pushing my hips in the air and my head down. I grab the line – a rope that is suspended from a buoy and anchored with weights that drop to the dive depth.

I pull myself deeper.

As I dive, a wetsuit clings to my body, forming a second skin that allows me to stay in the water longer without getting cold. Long fins morph my human legs into a mermaid-like tail. A mask fits tightly to my face, allowing me to see the line in front of me and the blue beyond. A rubber belt sits tightly on my hips. It is threaded with several kilograms of lead weights that help me sink, counteracting my body's buoyancy.

Though I could freedive without any of this gear, it transforms me, allowing for immersion that seems to transcend being human.

I began freediving while living on Lord Howe Island, far off the east coast of Australia. This verdant, volcanic island – home to about 400 people – is my research field site, where I am studying human–environment relations. Lord Howe and its Marine Park–protected waters are both listed as UNESCO World Heritage sites. Within these aquamarine waters lies the world's southernmost coral reef. 'This is our underwater playground,' says Liv Rose, my freediving teacher.

Before coming to Lord Howe, I had heard about freediving, but I thought it was a dangerous extreme sport I'd never try. That changed when I met Liv. I had always loved being in the ocean, and I took little convincing once I heard her talk about the unique feeling of freediving. Beyond that, I was anthropologically fascinated by why people might want to do something that seemed so counterintuitive: to hold their breath and fight all instincts to breathe in order to reach a new level of immersion.

Over the course of numerous classes, I absorbed the basics of freediving. Lying on the floor in Liv's house, I learned to breathe again. To breathe as a freediver does – fully – and then to hold it until I couldn't stand it any longer. With an oxygen monitor on my finger, I came to understand that though I felt I would black out any moment, I had 'so much time', as Liv would say in a calming voice.

I developed a new awareness of my body, discovering the parts of my anatomy I needed to control to equalise my ears. I researched people who have used the techniques of freediving for centuries. The Ama – Japanese fisherwomen divers. The Haenyeo – female divers of Jeju, South Korea, who harvest molluscs, seaweed, and other foods from the sea while holding their breath.

I learned about competitive freediving – the sport shown in the 2023 Netflix documentary *The Deepest Breath* and described in

James Nestor's book *Deep*. I wondered at the apparent obsession and addiction such freedivers have to get deeper, to push human boundaries.

My dive buddies and I don't go to such extreme depths. Rather, we are using the techniques to explore. And what we have found, in a sense, goes even deeper: a new identity and a realisation of what it means to be in reciprocity with the sea.

I am moving down slowly. At times I close my eyes to relax. There is no rush to the bottom.

Compared to scuba divers, who are laden with air tanks and take breaths that reverberate bubbles and noise, I move quietly, freely, gracefully. 'It makes you feel like you are from the sea,' Liv says.

At around 10 metres deep, I reach a point where my buoyancy cancels out. I neither sink nor float. Below that, the water starts pulling me down. At these depths, freedivers can enter a free fall.

I descend deeper.

Each metre down, the water pressure increases. It makes my lungs shrink, my airways squeeze. With my free hand, I clamp my nostrils together and blow, equalising the pressure in my ears.

I reach the bottom of the line.

I'm 20 metres deep.

I turn.

Tugging on the rope, I begin my ascent. This is just a taste of the addiction to depth. The freediving world record is 214 metres.

Without breath, carbon dioxide begins to build in my body.

My diaphragm tightens.

My ribs convulse.

These contractions come suddenly. They are my body's way of telling me I should breathe. I am still metres away from the surface.

I'm running out of air.

I calm these thoughts. I know I can push past this human

impulse. I've practised on dry land, holding my breath for three minutes. In the mindful state of freediving, I don't panic. I find stillness. Centeredness. Calm. I am belonging in the moment. I've retrained my mind to be underwater.

My body, meanwhile, is using the mammalian dive response to function. This is a set of physiological reactions that all mammals have, including humans. It's triggered by being in water, especially when your face is submerged and you're holding your breath. It slows your heart rate to increase your ability to preserve oxygen. (This is one reason it feels calming to splash water on your face.) Your body also shifts blood from your hands, feet, arms, and legs to your core, vital organs. This allows freedivers to stay longer in the state of apnoea – meaning 'without breath'.

As I near the surface, the urge to breathe builds. The contractions in my stomach become more severe. My dive buddy descends a few metres, meeting me on my return. She watches me for any signs that I might black out – which, underwater, can prove fatal.

We break the surface with a splash. I cling to the buoy. My dive buddy watches me intently. I could still black out. As Liv taught me, I take three exaggerated 'recovery' breaths – fast inhales and long exhales. I make an 'OK' signal to my buddy with my hand, saying the same. I've returned to the world above the surface.

One day, exploring underwater while freediving, Shawnee, one of the seasonal tourism workers, found an old car tyre wedged in the coral reef. She decided that the rubber, which is toxic when it breaks down in the sea, didn't belong. So she dragged it behind her on the swim back to the beach, then hauled it from the water for disposal.

Like many young people working seasonally in tourism on Lord Howe Island, Shawnee regularly bounces between places. I wondered, in my conversations with many of the seasonal workers, what might connect someone who moves so frequently to a sense of belonging.

'Underwater is where I feel truly accepted, at peace,' Shawnee said. 'Maybe that is what home means to me.'

Freediving – like surfing, swimming, and other forms of immersion – gives you the feeling of transcending being human. It dissolves our sense of separateness from the ocean ecosystem. It brings a feeling of belonging. A sense of what I call a 'citizenship of the sea'.

This relationship is transformative. It can translate to a desire to give back. Anthropologists call such exchanges of mutual benefit 'reciprocity'.

Seeing Shawnee pull a tyre from the reef and watching many others collect beach-washed plastic, or rescue birds who had ingested it, I came to understand these acts as a form of reciprocity between people and the sea. These demonstrations of care simultaneously preserve environments and help ensure that people now and in the future may have the same experiences and opportunities to find happiness and belonging underwater.

Today amid a growing rift between humans and nature, it is important to understand how we belong in environments. How we – humans and other species, land, sea and air – are all interconnected. By exploring new depths of immersion, enjoyment, and kinship in environments – through freediving, hiking, or other forms of being in nature – we can forge connections with the natural world and deeper relationships of care.

A sense of reciprocity with nature, above or below the sea, motivates us to give back more than we take. Through freediving, I've realised that when reciprocity flows from belonging, care is as necessary and natural as the desire to breathe.

✳ *Whale talk*, p. **159**

A museum heist 70 years ago is still causing a flutter in butterfly science today, p. **275**

FASTER HIGHER STRONGER DOPER

Matthew Ward Agius

What the hell?

There was no other way to react to the bizarre headlines that dropped in November 2023.

'Tour de France rider tried to obtain marine worm haemoglobin for blood doping boost,' read *Cycling News*. 'I never thought the next breakthrough in doping would be fishing worms,' wrote an op-ed from *Cycling Weekly*. More poetically: 'Opening a can of worms' from the bike journalism project *Escape Collective*. It certainly had.

Any mention of doping in cycling probably sends fans into a nail-biting chatter as they remember the halcyon years of 1990s juicing.

This story, though, is something far odder. We're not just talking about blood from a human. Not even another mammalian species; we're climbing the ladder of Linnaean classification – Genus, Family, Order, Class (looking back at all our fellow mammals), Phylum (saluting every animal with a backbone), and up to Kingdom, where we neatly jump over to the annelids (the segmented worms) and scramble all the way back down to the genus *Arenicola*. That's where oceanographer Franck Zal landed in the noughties, when he was based at the French national scientific research organisation CNRS and the Sorbonne University. His motivation wasn't juicing his favourite French pedallers.

Instead, his research had identified potential applications

for haemoglobin – that all-important protein responsible for transporting oxygen to our tissues – extracted from a species of European lugworm (*Arenicola marina*).

Worm haemoglobin is far more potent than the haemoglobin humans possess, and Zal saw its promise as a therapeutic that could support the preservation of transplanted organs. But that unnamed Tour de France athlete clearly saw the possibilities of harnessing worm blood to turbocharge his circulatory system for the big race. To understand why the frontier of performance enhancement has athletes looking elsewhere in the animal kingdom, it's important to consider where these biological boosters come from, and what they do.

Power of the protein

Almost every vertebrate contains haemoglobin proteins in their red blood cells. Its job is vital: delivering oxygen to tissues through the blood. In mammals, haemoglobin consists of four subunits, each a long, folded chain of amino acids that determine the protein's properties and function. These subunits are each connected to a heme group: a ring of organic compounds that contain a single iron ion. This iron binds with a single oxygen molecule and aids its transportation around the body.

With four subunits, each haemoglobin can carry four oxygen molecules. Think of haemoglobin as a four-seater car. The car itself is the haemoglobin, its four seats the subunits, and the driver and their three mates are the oxygen molecules being transported to their destination.

In other mammals, and in most vertebrates, haemoglobin is similarly structured – including having four seats for oxygen – and performs the same role. This consistency has enabled the development of new blood transfusion products from the haemoglobin of cows and pigs. Worm your way cross-Kingdom, and you'll find that *A. marina* haemoglobin performs the same role too.

Except it has not four, but 156 of these oxygen terminals: that's 39 times more carrying capacity than paltry human blood.

This advantage led Zal to start his own company – Hemarina – 15 years ago, with the goal of producing *A. marina*–based blood transfusion products at scale. The benefits are a clinician's dream: *A. marina* haemoglobin is a universal donor and appears to lack the side effects of other non-human and artificial sources such as in early haemoglobin-based oxygen carriers (HBOCs), which caused hypertension, vasoconstriction and oxidation.

The lack of side effects may be thanks to the way the molecule has evolved in worms: it is highly stable, resists oxidation and floats freely in the animal's bloodstream as opposed to being embedded within blood cells, as it is in vertebrates.

Mouse studies using common earthworm *Lumbricus terrestris* haemoglobin (with 144 oxygen terminals, by the way) have also shown an absence of adverse physiological responses.

Lugworm blood is now being adopted as a support mechanism for those in need of oxygen, especially organ donor recipients. Hemarina's HEMO2life™ product has been used to support organ preservation during an upper limb transplant in India and a facial transplant performed on a French soldier. The product was granted CE certification by the EU in 2022, enabling it to be sold across Europe.

But the big question for our potential blood-doping cyclist: how do they actually work?

Doping deep dive

Through training, skill development, a rigorous diet and sometimes a rare mix of genetic gifts, humans have pushed themselves to go faster, higher and stronger since the first Olympics. Records tumble every year, across every sport.

But where winning and losing are separated by wafer-thin

margins – sometimes requiring a photo finish – the dark arts of performance enhancement are seductive.

According to data from the World Anti-Doping Authority (WADA), nearly one in 80 athletes globally use performance-enhancing drugs, and one in every 215 Australian athletes. Those who take the bait seek an edge to run a little faster or lift a little more, using a chemical shortcut. Shortcut is the keyword here. There's no magic pill that transfigures a scrawny layabout into a medal-capable mega specimen.

Performance enhancement – whether substance or technique-based – simply enables the body to do more, work harder, or recover more quickly than would otherwise be possible.

Historically, there have been two favoured methods of performance enhancement. The first is anabolic steroids – once the drug of choice for athletes looking to build muscle mass and recover faster, and still in use today. Prohibited in and out of sanctioned competition, they are the most studied class of performance-enhancing drugs. In the simplest terms, these synthetic products assist the development of muscle mass. Whether a powerlifter, sprinter, rower or cyclist, an athlete's muscles need to be in peak condition to provide the strength and power needed to effectively perform their role. When used as part of a weight training program, anabolic steroids promote the growth of new muscle (thus giving athletes more force-generating ability), as well as aiding recovery by accelerating protein synthesis – effectively enabling the user to do more work, more often.

The second is the rise of blood doping from the 1980s onwards as a way to enhance aerobic capacity – this is where the promise of worm blood lies. How it works is remarkably simple.

Muscles have an enormous capacity for work, but our circulatory system is the limiting factor.

The performance of endurance athletes relies on their ability to use oxygen to produce energy. Oxygen is essential to cellular respiration, where glucose from food is subject to oxidation, resulting in the release of carbon dioxide (exhaled), water (sweat) and energy in the form of adenosine triphosphate.

All of this takes place in what's commonly called the 'powerhouse of the cell' – the mitochondria. But glucose and oxygen don't simply materialise there; they need to be ferried through the blood. Haemoglobin is transported by red blood cells, which are the body's cargo ships. But we only produce so many. Along with cardiac output (the amount of blood the heart is pumping around the body), the amount of oxygen in each litre of blood provides a performance ceiling.

Kenneth Graham, former principal scientist for the New South Wales Institute of Sport and now an anti-doping policy and research consultant, explains that expanding the fleet of red blood cells available to move oxygen is a handy way to boost an athlete's energy stocks.

'If we increase red blood cells and haemoglobin, we have more oxygen being transported per litre of blood around the body, we have a higher VO2 max [an individual's maximum oxygen capacity], we have a greater capacity to do aerobic work,' says Graham, who worked with many Australian Olympic and Commonwealth gold medallists between 1992 and 2020.

This increase can be done naturally through good diet, exercise and innovative training. Some athletes will, for instance, travel to higher altitudes, where the body adapts to less available oxygen by producing more haemoglobin and red blood cells.

Or you could just dope.

There are typically two ways to cheat your way to higher haemoglobin.

The first: extract the blood, centrifuge out and then reinfuse the plasma, and refrigerate or freeze the rest, to be reinfused before competition. Since your blood naturally replenishes the missing blood, an athlete will basically be injecting a bonus later on, like buying an iced coffee and dropping in an extra teaspoon of instant for a bonus caffeine hit. This is classic blood doping: an instant haemoglobin booster.

The second: inject yourself with hormones. EPO – erythropoietin – is the glycoprotein utilised in massive doping scandals like cycling's 1998 Festina Affair and by the US Postal Service team, led by Lance Armstrong, described by the US Anti-Doping Agency in 2012 as 'the most sophisticated, professionalised and successful doping program that [cycling] has ever seen'. EPO is naturally produced by the body's endocrine system in response to low blood oxygen.

'So you can reduce the training time between sessions because you've got an accelerated recovery,' Graham says. 'That, combined with more training, you get the contribution for an enhanced performance capacity.

'In the body, we have self-regulating mechanisms. EPO is produced in the kidneys, the renal medulla, in response to reduced oxygen levels ... The kidneys are basically saying, "We're getting less oxygen, we'll fix this up, we'll release EPO, it will go and cause the production of new red blood cells, which will increase the O_2 transport and the body will get the oxygen we need."'

When athletes head to higher altitudes, the hypoxic environment stimulates the kidneys to release EPO. This signals the bone marrow to produce more red blood cells, thus enabling blood oxygen levels to normalise. When they come back down the mountain, they're better equipped to tackle that next demanding event.

But EPO can also be unnaturally topped up through injections, much like the reinfusion of blood products. So, Graham says, 'even if the body shuts down its own production of EPO, it's still got this exogenous supply that's stimulating the production of red blood cells'.

But how do athletes get their hands on new substances to dope with?

'Many of the products that are used for performance enhancement have actually been derived from clinical use,' says Rhonda

Orr, director of movement sciences at Sydney University's School of Health Sciences. 'New drugs and processes have been developed because there's a clinical need,' she adds, giving the example of synthetic red blood cells, which were developed for people with severe anaemia.

Natural or synthetic EPO is also a vital therapeutic for people with damaged kidneys (as with chronic kidney disease) or some forms of blood cancer. Anabolic steroids can treat a range of hormonal issues such as delayed puberty in males, as well as encourage muscle growth in those suffering illnesses like cancer or HIV.

But advances in medicine can be nurtured, massaged and modified by scientists with other inclinations. Orr describes such scientists as 'enterprising chemists'. '[They] see these new developments and they can see the potential application to performance enhancement.' The Hemarina product is no different: developed to offer a human-compatible oxygen carrier to supplement limited blood stocks, it's a possible doping agent as well.

So with new and different products being taken from medical science and used for performance enhancement all the time, how do sporting bodies keep up?

Hunting the dopers

If, instead of going to Paris in mid-2024, you jumped in a time machine back to the first Olympics in Greece, you'd find a lot of naked athletes strutting their stuff in the ancient arena. You'd also find some of them, according to historians, trying to enhance their performance through the use of plants and fungi. Some scholars suggest that the use of these external substances wasn't discouraged.

Wind the clock towards the present day and the cases of performance enhancement begin to rack up. The first bans on stimulants were introduced in 1928; in 1960 a Danish cyclist who died at the 1960 Rome Olympics was found to have amphetamines in his system; in 1967, the International Olympic Committee listed the first banned substances. Tests for drug use were slowly introduced – once a substance is known, it's possible to develop a process to spot it. A notable moment in the early fight against drug cheating came at the 1988 Seoul Games, where Canadian sprinter Ben Johnson was stripped of his 100 metres gold after testing positive for the banned steroid stanozolol.

Then came the fall of the Berlin Wall, the reunification of Germany, and the opening of records from the former East Germany (GDR), confirming that drugs had systematically been administered to its athletes, helping catapult the communist state to sporting success in the 1970s and '80s.

Tests of athletes' blood and urine were conducted at approved labs – including ones in the GDR – but without a uniform body to oversee sample acquisition and analysis. That changed following the Festina Affair at the 1998 Tour de France, where evidence of a sophisticated EPO doping program was uncovered by police raids on that team's vehicles and hotel rooms. The World Anti-Doping Authority was subsequently established in 1999.

At the Sydney Olympics, the International Olympic Committee debuted a test to screen blood and urine samples for EPO use. Four years later, in Athens, a screen for human growth hormone was introduced.

Today, anti-doping authorities test athletes' biological samples for hundreds of banned substances, all inscribed on WADA's prohibited list, including anabolic agents, peptide hormones, beta-2 agonists, hormone and metabolic modulators, diuretics and masking agents, stimulants, narcotics, cannabinoids and glucocorticoids.

Banned methods are listed too, including the administration or reintroduction of any quantity of blood to the body, and the manipulation of the blood through physical or chemical means, as well as chemical or physical alteration of samples.

According to Mario Thevis, a biochemist who heads the

Centre for Preventative Doping Research at the German Sport University in Cologne, three main techniques are used by testing labs to find performance-enhancing drugs.

The first is mass spectrometry, where the molecular mass of a sample is measured to discern the presence of target analytes – profiles of banned chemicals. Immunological assays (tests) are performed to screen for the presence of other molecules, for example human growth hormone.

Electrophoresis (a process that separates molecules based on size and electrical charge) is also often used to find abnormal blood proteins. As red blood cells tend to reduce in size when extracted for storage and reinfusion, this technique can identify instances of blood transfusions of EPO.

Biochemists like Thevis don't just analyse athlete samples for current banned substances or methods: they also develop tests for new ones.

It's not easy to find substances previously unknown to tests. Both Thevis and Graham relate the story where an anonymous syringe arrived in the mail at the headquarters of the United States Anti-Doping Authority in 2003.

The clear liquid within was tested and retested, until eventually analytical chemists at the University of California Los Angeles cracked it: the syringe contained tetrahydrogestrinone, or THG – a steroid never made available for medical use.

Thevis says designer steroids like THG are modified from existing structures – just enough to retain their anabolic properties – 'but they were not immediately on our radar because they were slightly modified'.

Within three months, the California laboratories of the Bay Area Laboratory Co-operative (BALCO) were raided. Seized records shed light on who was taking THG: four track and field athletes during the 2003 national meets, as well as National Football League and Major League Baseball (MLB) players; one in

20 MLB tests were positive. US runner Marion Jones was stripped of her five Sydney Olympic medals after admitting to taking THG.

Without that syringe, it would have taken anti-doping authorities much longer to hook onto THG. But identifying a questionable chemical is not the only signal testers need. Knowing what the body does with it provides further clues.

'If you take a drug, a urine sample is collected and we analyse it. We can either target the substance you took or the bio-transformation product – the metabolite,' says Thevis. 'The entire drug might disappear entirely and we need to look for breakdown products.

'Once we know about a general structure of a new therapeutic class or a specific substance ... then we can start developing test methods. We do biotransformation experiments with cell cultures, or animal experiments, or, if it's a [clinically] approved drug, we collaborate with clinics where the drug is therapeutically administered to patients [and] we get approval to sample those patients to have authentic material to work with. Eventually, all we need to have is an analytical platform that allows us to identify those prohibited substances in blood or urine.'

Platform testing

So what about our lugworm blood? Is it a gold pass to an ill-gotten gold medal?

When French colleagues sent *A. marina* their way, Thevis and his team set about seeing whether it could be detected. They employed a liquid-chromatography-mass-spectroscopy-based method, which has been used for about two decades to detect doping with oxygen-carrying products like transfused haemo-globin and HBOCs. They modified this existing mass spectrom-etry test and injected Hemarina's HEMO2life^{TM} – containing 40 mg/mL of the active worm hemes – into three male lab rats.

'We saw that [lugworm haemoglobin] can be detected because its amino acid sequence is different from bovine and porcine and also from human haemoglobin,' Thevis says. 'With some minor modifications to our sample preparation procedure, we were able to include that new analyte into our testing platform with the information that it might have advantages over earlier first or second generation HBOCs.'

The results were encouraging. The test could detect 10 micrograms of lug heme per millilitre in a 50-microlitre sample. Lugworm haemoglobin doesn't last long. In rats it could be 'unambiguously detected' for 4–8 hours after administration, though in one sample traces were found after two days.

That, says Thevis, means it would only likely be used in competition, not training. Athletes using haemoglobin from fishing bait are rolling the dice. 'If you're tested in competition, a detection window of eight or even 12 to 24 hours for lugworm haemoglobin is probably sufficient, because that covers the most relevant period of the drug's assumed action on the athlete.'

But assuming that a best-case 48-hour window exists for testers to detect worm blood, a well-timed doper could plan their transfusion strategically. Say you have a three-hour marathon – you might get sampled immediately after the race, then it takes a few hours for the sample to be transported to a lab for testing, then it's tested, then verified. If you doped two days before, you could still derive some benefit from the short-lived annelid haem before it becomes undetectable by testing time.

Thevis hears that, and raises the Athlete Biological Passport (ABP). An ABP is a data catalogue of your biology, broken down into steroidal, haematological and hormonal modules. Every blood and urine sample informs and refines this profile over time.

Unlike lab tests, which hunt for specific banned substances or resulting metabolites, the passport works over extended periods to identify abnormal biomarker readings that last well after the act. Those stand out as red flags when compared to the rest of the profile, indicating the need for further investigation. Even blood donations – the ones you do every three months for the Red Cross – show up in the profile.

'That's a change that is accepted,' Thevis says. 'But if there's a change in the profile that indicates blood withdrawal and blood re-transfusion ... that will end in [disciplinary] proceedings.'

Keen to shake its reputation as the sport of dopers, professional cycling became the first sport to implement ABPs back in 2009. Other sports have followed, including running. But ABPs don't yet entirely protect athletes.

A strong passport

In January 2023, Australian runner Peter Bol was provisionally suspended from competition after an EPO test showed he had elevated levels as compared to his ABP baseline. A second sample, drawn to confirm the result, returned an 'atypical finding'. Bol's samples were then analysed by other accredited labs, and WADA experts were consulted. By August 2023, the first sample was deemed negative and Bol's name was cleared. It's since been suggested that Bol may have naturally elevated levels of EPO – a blessing for his athletic prospects, but a curse when a test flags it as a potential sign of doping.

'Hormones have really put those tests to the limit,' Orr says, 'because you can't just say "Let's just measure someone's growth hormone and, aha, it's higher than we expect – they must be doping!" Because hormones are endogenous [produced by the body] and everybody is so different – there's such variation in the human – they've had to come up with other tests.'

The solution to situations like Bol's could be to test early and test often, building up a time machine of biology that establishes an athlete's natural ranges, starting early on in their career. Frequent testing could also help pinpoint the window of time in which the change occurred: for example, if an athlete tested positive for a certain metabolite just two weeks after their last test, then authorities could more easily narrow down a potential cause – possibly an accidental diet change – within that window of time.

'As counterintuitive as it might sound to an athlete, the more you're tested, the less likely it is that you receive an anti-doping rule violation,' Thevis says. 'If you're tested with a tiny amount today, and your last test was negative, and your follow-up test is also negative, then in most instances you can't have had a pharmacologically relevant dose – or a doping dose – in between.'

But if there was a six-month gap between samples, plenty could happen naturally to an athlete's body chemistry – including changes to a training program, illness or altitude training – that may instead be flagged as suspicious.

'Overall, the more tests that these athletes do, the greater and more reliable their Athlete Biological Passport is going to be,' Orr says.

With that in mind, worm blood doesn't sound like the secret sauce to Paris gold. That misguided athlete digging up lugworms from their local beach would be better off putting the worms on the end of a hook instead.

✱   *Humanoid robots: Made in our image*, p. **12**
    Mirror molecules: The twisted problem of chemical chirality,
        p. **104**

INSECT CONSCIOUSNESS

Amalyah Hart

Dinis Gökaydin plucks a vial from the counter and holds it up to the light. Inside, a female fruit fly (*Drosophila melanogaster*) is crawling up its glass sides towards the false promise of freedom. Fruit flies do this when trapped – they travel upwards, buoyed by an instinct that tells them sky means salvation.

'It's a bad morning because I'm feeling shaky for some reason,' Gökaydin says. Nervous energy punctuates the air.

What follows is one of the most precise and nerve-wracking pieces of practical science I've ever seen. First, he places *Drosophila*'s tube into a small opening in a metal block on the bench in front of us. The inside of this block is 2 degrees Celcius, which, after three minutes, immobilises the fly.

Next, Gökaydin opens the vial and teases the fly out with a tiny instrument. He dabs some glue on the back of her body, and fastens her to a small metal plate, not much larger than a sim card. She's ready for the final, macabre insult.

Peering down a microscope, Gökaydin uses another instrument to peel open the back of her head, into which his supervisor Bruno van Swinderen will embed an electrode. I breathe a sigh of relief – he's managed to achieve all this without killing her, a crucial part of the plan.

Gökaydin and van Swinderen, a professor at the University of Queensland's Queensland Brain Institute (QBI), study how fruit fly brains function in different states – awake, asleep, or anaesthetised.

They hope their work may throw up clues to an enduring scientific mystery – how, and when, did consciousness evolve?

A short history of consciousness

Seventeenth-century philosopher René Descartes believed that consciousness was a uniquely human trait. Animals, he argued, made no effort to communicate their experiences. Plus they lacked the ability to reason. He therefore believed that, while animals might cry or run from pain, they had no thoughts or sensations – their behaviour was pure reflex.

This view became culturally widespread and, just a few decades ago, any attempt to investigate animal consciousness scientifically would have been derided.

In 1974, American philosopher Thomas Nagel published a provocative essay titled 'What is it like to be a bat?', which parsed out the problem of creature consciousness. Unlike Descartes, Nagel believed consciousness was widespread in the animal kingdom, but impossible to understand. To be conscious is a completely subjective experience, while science is governed by objective inquiry, Nagel argued.

Bats were the perfect poster child for this quandary: like us, they navigate the physical world, but unlike us, they do so through echolocation. Biohacking notwithstanding, most humans will probably never know how it feels to navigate the world through ricochets of sound.

Nagel had a point. The hunt for consciousness in the animal kingdom is hamstrung by the fact that no one can agree on what exactly consciousness is, or how it got here, let alone how to prove it's there.

Scientists remain divided on whether consciousness is a profound but accidental by-product of cognitive evolution, or a central driver of our species' evolutionary success. Some think it emerged as early as the invertebrates of the Cambrian explosion, more than

500 million years ago, while others think it appeared much later, in mammals and birds.

Numerous studies identify brain regions that may be involved, but there, too, disagreement reigns: is it housed in the front or the back of the brain? Is it in the neocortex, which only mammals have, or the brainstem, a region we share with most animals on Earth? And is consciousness tied to just one brain region, or made through the combined efforts of different neural circuits, thrumming together as one?

To make matters worse, there is no universally agreed upon definition of consciousness. Some researchers refer to 'consciousness', while others call it 'sentience', which refers specifically to the ability to feel and sense.

Other theorists divide it into 'cognitive consciousness', the ability to process information and solve problems, and 'phenomenal consciousness', which describes the capacity to have a subjective experience – to feel pain, see the colour red, taste the sweet flesh of a peach. Many scientists believe that subjective experience – this trait that, for us, gives life its richness – is not present in all animals.

Despite this lack of consensus, theories of consciousness are already pushing the needle on policy decisions. In 2022, the UK government moved to protect all vertebrates and some invertebrates, such as lobsters, crabs and octopuses, under its updated *Animal Welfare (Sentience) Act* 2022. That change was fuelled by a growing body of research that suggests these creatures may experience pain.

In 2009, Robert Elwood, a professor at Queen's University, Belfast, exposed hermit crabs to a series of electric shocks of gradually increasing intensity. Hermit crabs demonstrate clear preferences for certain species of shells, and Elwood wanted to understand how pain might influence their choices.

Crabs in higher-quality shells would suffer higher-intensity shocks than those living in lower-grade shells before they finally, reluctantly evacuated. What's more, when exposed to the shocks, crabs behaved in ways that looked achingly like true pain. Some crabs made escape bids, desperately trying to scale the walls of their tanks, while others furiously groomed the place on their body where the shock was administered.

'That's exactly the sort of profile of behaviour we would expect to see if crabs had a state that does for them the sort of thing pain does for us,' says Dr Jonathan Birch, a professor of philosophy at the London School of Economics (LSE), who was asked to review the evidence by the UK Government ahead of the 2022 change to the Act.

'Admittedly this is an area where there's a huge amount of uncertainty,' adds Birch, who heads the Foundations of Animal Sentience (ASENT) project at LSE. 'But I think the evidence is pushing us to take seriously a realistic possibility of conscious experience being extremely widespread.'

The oddball paradigm

The fly is suspended in front of an LED screen, which is divided in half. Every few seconds, a panel of light will appear on either side. The light has an equal probability of appearing on the left or right side each time – a computer generates the patterns at random. Meanwhile, the electrode in *Drosophila*'s brain is registering her reaction to each flash – unusual patterns create pulses of electrical activity in her brain.

Gökaydin and van Swinderen investigate attention, sleep and memory in fruit flies, and van Swinderen has been probing fruit fly brains for the better part of two decades. About a decade ago, van Swinderen set about designing a system that could monitor *Drosophila* brains while they slept. He found that in fruit flies, as in humans, sleep consists of both active and quiet phases. In humans, our active phase of sleep is known as REM sleep – the kingdom of dreams. Van Swinderen's data forced him to take seriously the possibility that fruit flies, too, may dream.

But today's experiment requires this fly to be wide awake. It's based on a psychological concept called the 'oddball paradigm'. When the brain encounters a novel stimulus – an 'oddball' – it experiences a spike in activity as it attempts to understand what's going on, and assess for threat. That spike is connected to the feeling of surprise; in humans, surprise is a jolt that funnels our awareness to the stimulus. But surprise is also costly.

'If you have a trillion synapses in your brain, even if you increase the synaptic release by one per cent, you're looking at a tremendous amount of extra energy requirements,' van Swinderen says.

So, brains need to optimise their function while minimising wasted energy: 'One way to do that is to be predictive, rather than reactive.'

Van Swinderen believes consciousness is an evolutionary adaptation that helps animals make predictions about the world around them, by focusing attention on difficult problems.

'I would really be of the opinion that consciousness is adaptive,' he says. 'If you had a simple animal that, within its very limited environment, was making 100 per cent perfect predictions, it wouldn't need it.'

Van Swinderen and Gökaydin aren't the only ones interested in fruit fly cognition and behaviour. Another 2021 *Drosophila* experiment, from a US-based team of researchers, found that chronic social isolation interrupted fruit flies' sleep cycles, and led them to overfeed – a phenomenon linked to loneliness in humans.

It's important to note that data that hints at loneliness, or dream-like states, doesn't prove that these experiences are anything like ours, or that these creatures are necessarily having an 'experience' at all.

Van Swinderen himself is not certain that insects are conscious in a way we might relate to. But he sees *Drosophila* as a chance to investigate the origins of subjective experience in the animal kingdom – there's no fire without a spark.

The psyche of bees

'You might want to tie back your hair,' says Andrew Barron, a neuroethologist at Macquarie University's Minds and Intelligences Initiative in Sydney. Neuroethologists study the neural mechanisms of animal behaviour.

It's a month after my visit to Brisbane, and this time I'm chasing the psyche of an altogether different insect.

'Don't worry, we've always got EpiPens onsite,' adds Barron's PhD student, Théo, who strikes me as someone who has felt the keenness of a bee sting many times before.

I've pulled my hair up and under a giant beekeeper's hat, my head and shoulders veiled. But my arms are bare on this sweltering January day, a fact about which no one seems particularly concerned.

The metal gate swings open and the three of us step into a giant, house-sized cage, filled with thousands of honeybees. Barron asks me not to step any further, because this particular hive is a little 'spiky'. The lexicon suggests that hives, like people, have minds of their own – and this one has a short temper.

At the near end of the cage, Dr Marie Geneviève Guiraud, a researcher at the Institute, has set up a small white box on a trestle table. Inside the box is a miniature arena, in which she tests whether bees can distinguish between different human faces, rewarding correct answers with sugar water. (Spoiler alert: they can.)

Next to the box, a single bee is sitting patiently, a small, square chip fixed to her abdomen. This is the bee Guiraud has been training, and she tells me she'll often find her there waiting, obedient as a trained puppy.

In the insect world, bees have a monopoly on charisma – famously intelligent, admirably cooperative, and cheerfully patterned. Along with *Drosophila*, they're also some of the most widely studied insects, in part because their cooperative nature and workhorse mentality makes them easy to train.

Bees possess a particularly flexible intelligence. They can plan and remember complex routes, and make rapid, accurate decisions as they drift from flower to flower. Bee cognition is being studied to inform robotics and AI.

There is also evidence that bees may have some basic emotional states, such as the capacity for optimism and pessimism.

In one 2011 study, scientists agitated honeybees by vigorously shaking their containers to simulate a predatory attack. Agitated bees were more likely to predict a negative outcome when exposed to an ambiguous stimulus than unmolested honeybees exposed to the same stimulus. The agitated bees also displayed lower levels of dopamine and serotonin, neurotransmitters associated with pleasure and happiness in humans.

'I think it's at least reasonably likely that bees and some other insects are conscious,' says Lars Chittka, a neuroethologist at Queen Mary University (QMU), London, and author of the book *The Mind of a Bee*, though he acknowledges it's impossible to prove.

A 2022 experiment from Chittka's team at QMU showed that bumblebees will go out of their way to roll wooden balls around, despite no obvious incentive for doing so, with all the hallmarks of playfulness. And younger bees roll the ball more often than older bees, just as older mammals play less than their young.

'They return to this activity again and again, when there's no reward present, so that's a hint that they enjoy the activity itself,' he says.

In 2016, Barron, who heads up the Macquarie Initiative, co-wrote a paper with Colin Klein, now a philosopher of neuroscience at the Australian National University, arguing that insects may well possess at least a basic form of subjective experience.

Their argument centred on the midbrain, a small part of the brainstem at the very centre of the brain. This handy part takes in sensory feedback from the outside world, and uses it to create an internal simulation of an animal's position in space.

Barron and Klein argue that this representation of the world, as a physical space through which a 'self' moves, is enough to make subjective awareness – and structures in the tiny insect brain perform a similar function.

Other ideas of consciousness

'While the majority of philosophers and scientists think [phenomenal consciousness] is relatively primitive – in other words, that's where consciousness began and other computational, intellectual forms of consciousness arrived later – I think it's the other way around,' says Nicholas Humphrey, a renowned English neuropsychologist who has been working away at the problem of consciousness for most of his life.

In the 1960s Humphrey, then a talented young PhD, began working with a lab monkey called Helen. Helen had had her visual cortex removed – the part of the brain that, in mammals, is responsible for processing visual information from the eyes. She could distinguish between light and dark, but could not seem to see shapes or distance.

In mammals and birds, the visual cortex is the main brain region involved in sight, but we also share a second, more ancient visual pathway with other animals such as fish, reptiles and amphibians. In most animals, this pathway travels from the eyes to a region of the brain called the optic tectum; in mammals, it travels to the superior colliculus, the optic tectum's evolutionary descendant.

Helen's superior colliculus was intact and, through hours of observation, Humphrey suspected something interesting was going on with her vision. While Helen behaved as if she were blind – staring into space, colliding with objects, and moving cautiously in new spaces – she would sometimes reach for items Humphrey placed in her view.

Over several years, Humphrey trained Helen to negotiate obstacles, navigate the room, reach for fruit and nuts, and climb trees. When she was nervous or overwhelmed, her newfound visual abilities seemed to vanish, and she would behave as though blind again.

What Helen was living with was a neurological phenomenon we now know as 'blindsight', which also occurs in humans. Human blindsight patients who have damage to the visual cortex believe they cannot see – they have no conscious sensation of seeing. But, when asked to locate or identify an object placed within their visual field, they show much higher accuracy than if they were simply guessing.

For Humphrey, blindsight began to seed an idea about consciousness. Was it possible that animals like frogs were 'seeing' like Helen? Was it possible to have perceptions without sensations – to see without seeing?

If animals with an optic tectum but no visual cortex are capable of sight, but lack the corresponding sense of seeing, the same might be true for other sensations. In some animals, sensory information, like other major bodily processes, might exist solely in the realm of the subconscious. If true, those creatures would lack sensations altogether.

Humphrey believes phenomenal consciousness – the rich, vibrant experience that sensations give – is a recent adaptation present only in birds and mammals, and that it evolved to help those creatures negotiate complex social environments. Knowing your own inner world is crucial to operating in what Humphrey calls the 'society of selves'.

'I think phenomenal consciousness is a sophisticated brain operation,' Humphrey says.

When it comes to other creatures, including lobsters, crabs, fruit flies and bees, Humphrey believes they 'just don't have this feeling, living in the present tense of sensation – they have a more robotic consciousness'.

But Chittka believes the accretion of pieces of evidence that hint at emotion and sensation in insects are unlikely, taken together, to be an accident.

'If you were asking this question about robots or computer programs, then yes, you can get them to pretend to be sentient,' says Chittka. 'But I don't think that nature has room for the kind of profligacy to generate beings that just pretend they feel something.'

On pain and suffering

When the experiment is over, the fruit fly's prospects of a normal life are nil, so Gökaydin will euthanise her in the most humane way he knows how: the squish method.

Australia does not publish annual figures for the number of animals used in lab research, but global estimates suggest nearly 200 million creatures are tested on in labs worldwide each year.

The term animal here is not all-inclusive. Biologically, the animal kingdom encompasses all multicellular, eukaryotic organisms that consume, reproduce and breathe oxygen, but UK law, for example, only classes vertebrates and cephalopods as animals, based on this contested concept of sentience. That means that universities and research centres try to use 'lower' species like worms and insects wherever possible.

Drosophila, in particular, are a darling of the research world. They live fast, die young, reproduce prolifically and have just four chromosome pairs, making them simple creatures to study. Countless numbers of them have given their lives to science: six Nobel Prizes have been awarded to *Drosophila* scientists alone.

Meanwhile, around 23 billion animals are factory farmed annually, and the burgeoning insect protein market is forecast to reach US$9 billion by 2030, to feed a growing population in the face of climate change and biodiversity collapse. That's not to mention the gruesome toll of pesticides, which lead to a harrowing death for insects.

Science and agriculture both have the unenviable task of balancing their ethical responsibilities to humans and to animals. Since the science of animal consciousness is so unsettled, most invertebrates remain unprotected by welfare legislation. Jeff Sebo, a philosopher at New York University, would like that to change. 'My view is that as long as a being has a non-trivial chance of being conscious, we should give them at least some consideration when making decisions that affect them,' he says. 'A one per cent chance that insects are conscious means a one per cent chance that we might be causing trillions of insects per year pain and suffering.'

Both Sebo and Birch were among the authors and signatories of the New York Declaration on Animal Consciousness. That statement, published by a group of philosophers and scientists in April 2024, argued that evidence for consciousness was widespread in the animal kingdom, and may well extend to insects.

For Birch, our failure to empathise with insects is really a failure of imagination.

'We struggle to engage with them as a fellow sentient being, and I'm sure the feeling is mutual,' he says. 'We're this giant, looming presence in their lives, as inscrutable to them as they are inscrutable to us.'

But Birch is hopeful that a mounting body of evidence will give policymakers pause.

'There's this sea change in the culture of science, and people are now daring to ask questions about animal consciousness,' Birch says. 'We hope this will send a signal to policymakers that they cannot simply ignore the interests of animals completely, because they're capable of suffering.'

✱ *Why can't we remember our lives as babies or toddlers?*, p. **204**
From hypnotised to heretic: Immunising society against misinformation, p. **228**

AIR CONDITIONING QUIETLY CHANGED AUSTRALIAN LIFE IN JUST A FEW DECADES

James Purtill

Over the past three decades, one technology has snuck into the very centre of Australian life, shaping where and how we live. It influenced the design of our houses and offices, the cars we drive, and the length of our daily commute. It even changed our notion of comfort. It's everywhere, yet largely out of sight. It's plonked on rooftops, tacked to back walls, and piped under panels and upholstery.

Now ubiquitous, only a generation ago air conditioning was sold as a pricey extra, and a generation further back it was a rare engineering marvel.

So how did 'manufactured air' become a must-have? How has it already changed us – and how will it continue to shape our existence?

How air con went mainstream

Most years, around September, a little-known report is published on the Australian Government Department of Climate Change, Energy, Environment and Water website. Called Cold Hard Facts, it's mostly unread outside of refrigeration and air conditioning industries. But buried in its tables of dry figures is the story of how Australians – who broadly pride themselves on their ability to handle the heat – embraced air conditioning over a few short decades.

The early 2000s in particular saw air conditioning become the norm. In 1999, about 35 per cent of Australian homes had air con. By 2010, it was 70 per cent.

We can also see the steep rise in the number of air conditioning units installed each year. In 2000, it was 330 000. In 2022, it was 1.5 million.

These figures describe not only the arrival of air conditioning, but also the departure of a former way of life. A much hotter one. It was a world of open flywire windows, breeze-catching louvres, sweating crowds in pubs, creaky ceiling fans, and office workers in short-sleeved shirts and straw fedoras.

'People back in the day used to just accept that they didn't have air conditioning and were quite comfortable without it,' says Graeme Dewerson, principal consultant with Expert Group, which compiles the Cold Hard Facts report. 'We've really changed, haven't we?'

Seventy years ago, newspaper stories about 'weather con-ditioners', 'climate controllers' or 'manufactured air' were treated with incredulity. When Queen Elizabeth II visited Australia in 1954, air conditioning in cars meant the 'four and 60': four windows down at 60 miles an hour.

But this would not befit a young queen.

A Melbourne taxi-driver-turned-mechanic retrofitted the regent's fleet of British-made cars with early US-imported mobile air conditioning units. Newspapers described it as 'special equipment' that could lower the inside temperature. Only a few buildings at the time had air con.

Then, in the late 1950s, Australia's first 'skyscraper' rose in central Melbourne. ICI House was made of glass and steel, like a garden greenhouse. But instead of roasting its office workers alive each summer, it cooled them with conditioned air.

'You wouldn't be able to have the ICI building without air con,' Alan Pears, a sustainability expert and senior industry fellow at RMIT University, says. 'And the whole emergence of the high-density high-rise could not happen without air con.'

At the time, the 20-storey ICI House was a glinting monument to the power of air conditioning to change how and where Australians lived and worked.

From the 1960s, air-conditioned skyscrapers rose in city centres. Apartment towers were built taller. Yet the whirr of air conditioning units was rarely heard in the suburbs.

Air con conquers the suburbs

Through the '70s and into the '80s, most new cars in Australia were sold without air con, and most houses relied on what would later be called 'passive solar' design. Long eaves shaded the windows and big sleep-out verandahs caught cool breezes. A container of water below a fan provided some evaporative cooling.

This all changed when cheaper car and home air conditioning units hit the market. Suburbs sprawled further, often away from cool sea or river breezes, and air con made those hotter inland areas more habitable.

The arrival of cheap split-system residential air con units in the 1990s saw a turn away from traditional passive-cooling housing design. 'Air conditioning made building designers and architects quite lazy,' Sam Ringwaldt, CEO of Australian air conditioning company Conry Tech, says. 'You could literally build a box in the desert and as long as you put enough air conditioning in it, it can be a comfortable work environment.'

Air con in vehicles also made longer commutes possible, and by the 1990s it came standard in new cars. These days, people routinely travel in bubbles of air con between other, larger bubbles of air con.

The rise of air conditioning has even changed what we consider to be a 'comfortable' temperature. 'Traditionally, the building code designers used to assume you would tolerate up to 26 to 28 degrees Celsius before you turned on any cooling. Now we're finding lots of people will turn on AC when it's a fair bit lower than that,' RMIT University's Mr Pears says. 'It's meant that air conditioning has become integral.'

A warming world means more air con

Staying cool comes with a catch: it chews through a lot of energy. On the hottest days, the power grid almost exclusively feeds electricity to meet air conditioning's demand. In a hot and humid city like Brisbane, air con accounts for 90 per cent of household electricity use during peak times.

But air conditioning's underlying technology hasn't changed much since its invention more than a century ago.

An air conditioner is a compressor combined with a long closed-loop coil of refrigerant gas (a chemical good at absorbing heat). When hot air flows over the cold, low-pressure coils, the refrigerant inside changes from a liquid to a gas, absorbing heat in the process. To keep cooling efficiently, the air conditioner must compress the refrigerant gas back to a liquid again. This work of compressing refrigerant gases requires a lot of electricity.

Refrigeration (which relies on the same underlying technology) and air conditioning use a quarter of all electricity generated in Australia, according to Cold Hard Facts. The same report says air con accounts for about 12 per cent of total greenhouse gas emissions, or about the same as the national fleet of 20 million cars, utes and vans.

And that's just Australia. At a global scale, the amount of air con being installed is hard to comprehend. 'Around the world today, there's about 250 new air conditioner units getting installed every second,' Conry Tech's Ringwaldt says. 'And it will be that way from now till 2050. These are staggering numbers that people are unaware of.'

To put this in context, Australia currently has about 18 million air conditioning units. Based on Mr Ringwaldt's figures, this number of units is being installed globally every 20 hours.

And as temperatures rise, installation rates are going up accordingly.

'You've got developing nations like Indonesia, China, India, Brazil, that are gobbling up air conditioning, because these are some of the hardest hit places by climate change,' Ringwaldt says. 'They know they're getting hotter. Therefore they need air conditioning. And this helps make the planet hotter.'

More air con means more electricity use, further increasing the greenhouse gas emissions that drive global warming (most electricity globally is still generated by burning fossil fuels). Global energy demand from air conditioners will triple by 2050 on current trends, according to the International Energy Agency.

Recent international efforts have aimed to reduce air con's emissions. In late 2023, more than 60 countries signed the Global Cooling Pledge to cut emissions from cooling systems and increase air conditioning efficiency.

Australia declined to be a signatory.

Greenhouse gas emissions from old units

Aside from energy use, staying cool comes with another and lesser-known catch. Many of the refrigerant gases air con units need are very effective at trapping heat in the atmosphere. A standard wall-hung split air conditioning system may contain about 1.7 kilograms of the refrigerant gas HCFC-22.

HCFC-22 is a greenhouse gas hundreds of times more potent than CO_2. So a standard air con unit may contain the global warming equivalent of a few tonnes of CO_2.

Australia's 18 million air conditioners and heat pumps hold gases equivalent to 100 million tonnes of CO_2, according to Cold

Hard Facts. So long as these gases don't escape, they don't add to global warming.

Under state and federal laws, when an air conditioning unit is decommissioned, the gases are to be collected and incinerated to avoid dangerous emissions. But these laws are widely ignored, Graeme Dewerson says. 'There's around a 20 per cent end-of-life capture rate [of the gases],' he says.

That is, of the hundreds of thousands of air con units that are decommissioned each year by demolition teams, car wreckers, electricians and others in the broader industry, most of the time their refrigerant gases are not captured. This equals roughly 7 million tonnes of CO_2 every year released into the atmosphere, which is equivalent to the annual emissions from about 1.6 million petrol cars.

It's a 'terrible' situation, says Mark Mitchell, managing director of air conditioning company SuperCool, who's been in the industry for four decades. 'So many sectors of the industry are not prepared to coordinate and focus on the goal of making things better for the environment.'

Our lives are being quietly and unobtrusively shaped by a technology that we barely notice. Now, as the climate changes, we're forced to take a closer look at the machinery that maintains our comfort.

It's quite a sight. No wonder we're feeling sweaty.

✳ *The night I accidentally became a corpse flower's bedside manservant*, p. **9**
The world has been its hottest on record for ten months straight. Scientists can't fully explain why, p. **265**

MIRROR MOLECULES: THE TWISTED PROBLEM OF CHEMICAL CHIRALITY

Ellen Phiddian

When British skier Alain Baxter won a bronze medal, it was a personal best for him – and a first for the UK in skiing. Baxter came third in the slalom at the 2002 Salt Lake City Winter Olympics.

But then his urine sample tested positive for methamphetamine. Three weeks later, the International Olympic Committee (IOC) had stripped him of the medal.

Baxter denied taking the illegal stimulant, saying the culprit was a Vicks inhaler he'd bought in Utah to clear his nose. Unbeknown to him, the brand used a different substance in their US product to the UK version. While he could use the UK inhaler with impunity, the US formulation triggered a positive drug test.

The Court of Arbitration in Sport later cleared Baxter of cheating, ruling he 'appears to be a sincere and honest man who did not intend to obtain a competitive advantage in the race'.

But he never got his medal back. The court agreed with the IOC's ruling that, whether or not he'd known he was doing it, Baxter had still used a banned substance.

The thing is, the substance inside the Vicks inhaler had an identical chemical formula to methamphetamine. It had exactly the same atoms as methamphetamine, arranged in exactly the same chemical bonds. All of the IOC-accredited equipment in the Salt Lake City lab would have agreed it was methamphetamine.

But it wasn't methamphetamine. It wouldn't have given Baxter an advantage. In the US, you can still buy it over the counter in a number of inhaler brands (although Vicks themselves have since changed their recipe). Taking those inhalers won't cause a high, or any of the addictive effects of methamphetamine. They're recommended for children as young as six.

The Vicks ingredient was an almost indistinguishable twin of methamphetamine: R-methamphetamine.

The difference? A simple piece of geometry. A deeply vexing problem. An unprecedented threat.

Twin molecules

At one crucial point in the methamphetamine molecule, a single carbon atom links four completely different sections together.

This introduces a source of asymmetry, called 'chirality', into the molecule. Try and connect these four sections to the central atom, and you'll discover they can be connected in two different ways, with one the mirror image of the other.

The molecules end up looking like a pair of hands: made of the same stuff, connected in the same way, but not interchangeable. In fact, chemists term these reflections the S form and the R form in reference to handedness – S stands for 'sinister', Latin for left-handed, while R is 'rectus' (right-handed).

It's not a distinction that's easy to spot, chemically speaking. The molecules have all the same atoms and all the same bonds – they're just optically different.

Given molecules are too small to see, why should optics matter? Turns out, life is full of chiral molecules. Most importantly, both DNA and 21 of the 22 amino acids that make up all the proteins in living things are chiral – only R-DNA, and mostly S-amino acids, exist naturally. So when a human encounters a chiral molecule, its symmetry suddenly matters very much.

'No one knows why our bodies got only one form of amino

acids,' says Professor Ashraf Ghanem, a chemist at the University of Canberra.

While S-methamphetamine binds neatly to protein receptors in the brain, triggering the release of dopamine and other transmitters, R-methamphetamine does not.

Chemists term these symmetry twins enantiomers, or stereo-isomers. And methamphetamine is far from the only substance to have enantiomers that behave differently.

Mirror effects in the body

'There's a range of different effects if you have the wrong stereo-isomer,' says Associate Professor Vinh Nguyen, a chemist at the University of New South Wales.

Take carvone molecules, which are found naturally in essential oils. They carry distinct scents – spearmint from one enantiomer, caraway from the other.

But when it comes to medicines, the difference can be much more frightening. The vast majority of pharmaceuticals are molecules designed to fit into the 'active sites' of certain proteins in the human body, and these active sites are highly specific.

'The drug will be able to treat the disease because it was able to fit nicely into the active site,' says Ghanem. 'If it's not the correct form, then it will not fit, and the drug will be inactive.'

Add the wrong enantiomer to your medicine, and the best-case scenario is that it does nothing. Ibuprofen is an example of this: it's sold as a 50-50 mixture of both enantiomers, one of which (the S-enantiomer) is 100 times better at relieving pain than the other. It's wasteful – manufacturers are using twice as many ingredients to make the right dose of medicine – but for now, it's cheaper to make the mixture than it is to make or separate pure S-ibuprofen.

But that's the least harmful option.

'There's several cases in the history of drug development

where the wrong stereoisomer had very different effects, sometimes severe,' says Nguyen.

The most famous example is thalidomide, which was used to treat morning sickness in the 1950s and early '60s. One form of the drug – R – has a sedative effect. But in the human body, it converted easily into the S-form, which caused disorders in developing foetuses. About 10 000 infants were born worldwide with severe limb malformation, only half of whom survived.

'Before launching any drug, now, the TGA asks companies manufacturing drugs to show us the effect of each form separately and convince us that the drug is safe in terms of chirality,' says Ghanem.

So unless you want to do twice the amount of clinical work proving both enantiomers are safe, you need to make something enantiomerically pure. Herein lies the next problem.

Nature discriminates when it comes to chiral molecules, but chemists find it much, much harder. It's not a random coincidence that it's cheaper to make a mixture of ibuprofen enantiomers. That's the default result of a chemical reaction. Nature, which has spent millions of years reinforcing its chiral purity, is the weird aberration.

If you start with symmetrical feedstocks, how can you impose asymmetry on them? How do you get the right optical property in something that's too small to see?

'It has been a challenge in organic synthesis for a long time,' says Nguyen.

Pure asymmetry

One way to get a pure enantiomer is to make a mixture of both forms, and separate them. A common way to do this is chromatography: filling a tube with a chiral substance that will stick to only one enantiomer, and flushing the mixture through it.

'That will cost a lot of operating time, solvents and materials,' says Nguyen.

This technique can also help to identify enantiomers. In 2002, the IOC had access to chiral chromatography equipment at a Californian lab, but they denied Baxter's request to use it on his sample.

A neater way is to make something chirally pure in the first place.

'From the beginning, you can use a catalyst that can differentiate between R and S, and you will make only one of the drugs,' says Ghanem.

Start with a catalyst – something that triggers a chemical reaction, without being used up itself – and you can turn a tiny bit of chirality into a lot of chirality. Ghanem and colleagues have seen commercial success with this technique, making a catalyst that yields Ritalin. The catalyst – under patent – is now in the hands of multinational company Merck.

Historically, chemists have turned to transition metals like palladium to form the basis of these catalysts. The metal atoms have the right sort of structure to accommodate other molecules coming together and reacting in their presence. But nature's provided a carbon-based solution: enzymes.

'Enzymes are these really big proteins, with a lot of hydrophobic [water-repelling] pockets and some sort of active site to catalyse reactions,' says Nguyen. 'The reagents have to get inside that pocket, and then bind to those active sites, to react with each other better.'

Ghanem and colleagues have used enzymes to get to the chiral molecules they want.

Nguyen's team, meanwhile, is interested in combining the best of both worlds with a technique called organocatalysis. It relies on small, carbon-based molecules that can mimic the active sites of enzymes and the structure of metals to trigger reactions.

'Essentially, simplify everything and use what we know about the other catalytic interactions to bring it together onto this small organic framework,' says Nguyen. 'It uses all the advantages of the other two fields. But then [it doesn't] have to deal with bulky, complicated molecules, or transition metals, which are kind of toxic and hard to handle – and expensive, sometimes.'

A new kid on the block at only a few decades old, organocatalysis promises to be safer and more efficient than both its predecessors. Many of the catalysts used are derived from nature themselves.

'A bunch of the original organocatalysts are actually amino acids,' says Nguyen. 'So they have stereochemistry already built in, and it's quite easy to isolate them and use them to then coordinate or interact with other organic molecules.'

Nguyen's interested in making all of this synthesis greener – using sunlight as a source of power to trigger reactions, and figuring out ways to make the reactions happen in water. Carbon-based molecules generally require polluting and expensive carbon-based solvents to dissolve properly, so water-based reactions could clean industry up dramatically.

'There's a lot of things that we can use from nature, but human creativity is unlimited,' says Nguyen.

The risk of an evil twin

There's nothing inherently life-giving about a left-handed protein. The S designation itself is an arbitrary one, dreamt up by mid-century chemists, just so they could distinguish between enantiomers without resorting to drawing diagrams. There have even been boons to medicine from reversing natural chirality, and making right-handed proteins or left-handed DNA.

'These mirror molecules are really intriguing as a new category of potential drugs, because they are not recognised by the body – by degradative enzymes or the immune system,' says Professor Michael Kay, a biochemist at the University of Utah in the United States. '[Mirror] drugs could have very desirable properties in terms of very long half-life, low cost, infrequent administration – things like that.'

But Kay, who does research on these therapeutic drugs, is

concerned the work could go too far. In December, he was part of an interdisciplinary group of researchers who co-authored a stark warning in *Science*. Their concern was the possible creation of 'mirror life' – entire organisms created from the opposite enantiomers to ordinary life. It's feasible, the researchers suggested, to make bacteria with right-handed amino acids and left-handed DNA.

We can't make mirror bacteria yet. Every chiral part of the organism would need to be made from scratch, and the researchers think we're at least a decade of hard work away from such a feat.

'This is not an imminent threat. It's not science fiction either,' says Kay. The problem is that if mirror bacteria escaped the lab, it could make all previous pandemics look minor.

Most of the nutrients bacteria live on aren't chiral, so mirror life would have no problem replicating. But the things that control bacterial numbers – phages, antibiotics, other microbes, immune system proteins – they're all chiral. 'Since almost all of the immune system has this chiral requirement, you would have something that could escape many different aspects of immunity all at one time,' says Kay.

The effect would be deleterious – not just to humans, but to every living organism. The world would have no natural defences against these bacteria, and the researchers are sceptical that we could build artificial defences strong enough.

Mirror life should never be made, concluded Kay and his colleagues – any pharmaceutical or scientific benefits are not worth the risk.

The good news is that there's plenty of time to put rules in place, so that mirror proteins and other molecules can be studied, without any risk someone might make a whole bacterium. The group is inviting wide discussions this year to start developing these rules.

'This is not something that somebody could work on in their garage, away from a regulatory infrastructure,' says Kay.

Nor are they totally slamming the door. 'We do want to be humble in that we don't know everything,' says Kay. It's possible – albeit very unlikely, he thinks – that some new insight could render mirror life safer.

But chirality has been a source of medical disasters in the past, and for now, all the evidence suggests that mirror bacteria would be far worse.

Chirality was a trivial matter for the IOC, which denied Baxter's medal on the grounds that it had banned all methamphetamine. But for chemists, the distinction matters. An R or an S could be the difference between life and death.

We are getting better and better at choosing the symmetries of molecules. Will these choices lead to something beneficial – or sinister? For now, it's not quite clear. We see in a mirror – dimly.

✱ *Faster higher stronger doper*, p. **74**
   Some psychedelic medicine developers want to ditch the therapy aspect. What could go wrong?, p. **120**

HOW SURVIVING A DEADLY ANIMAL ENCOUNTER IS NOT THE END OF THE TRAUMA

Angela Heathcote

Paz Moreno hesitantly takes a step towards the water's edge at Chinamans Beach.

Located in the affluent suburb of Mosman on the shores of Sydney Harbour, there's little to be afraid of – the waves are small and avoidable, the turquoise depths nicely buffered by a comfortable stretch of sandy shore to safely plant your feet.

This was once one of Paz's favourite places to swim.

When she first moved to Australia from Chile in 2018, Paz quickly and eagerly adapted to Australia's beach lifestyle.

'Here, you use the beach as a part of your life every day ... I see people go to the beach in the winter ... there's no problem with that – it's just routine,' she says. 'In Chile, it's not so common to be in touch with ocean life.'

Despite this, she developed a reverence for the octopus – pulpo, in Spanish – from a young age.

'I felt very ... not scared ... but, "Oh we have to be careful with these animals", more than sharks or other kinds of creatures, because they are very intelligent,' Paz says.

For this reason, she 'never wanted to have an encounter with an octopus'.

Paz stands hand-in-hand with her partner Mauricio Quilpatay, their feet buried in the sand, ankle deep.

She dreams of going further – back into the water.

'I feel it will be soon. A moment in my life that I will return, but I don't think it will be like it used to,' Paz says. 'The ocean is so huge. It's a fantasy that we are in control.'

One, two, three pinches

In March 2023, just as the summer heat was fading, Paz decided to take a quick dip at Chinamans Beach with Mauricio, right before she taught her usual Thursday Spanish class.

She zipped up her swimsuit, adjusted her snorkel and entered the glittery blue water.

'I submerged, and I saw a shell. I picked it up and returned to the surface to look at it,' she says. 'I checked the shell because the shells can have a crab in it or a snail.'

It was seemingly empty, so Paz placed the shell in the pocket of her swimsuit, and continued swimming.

The first pinch came when she began to leave the water.

'It wasn't painful but was kind of annoying.'

A second pinch.

Paz checked her swimsuit and saw nothing.

A third pinch.

She noticed a lump protruding underneath her swimsuit. She held the lump and unzipped for another check.

There was the octopus, 'with the full bright blue lines'.

Paz was aware it was a blue-ringed octopus, as she'd recently watched a TikTok video about the dangerous animal, 'so it was pretty fresh'.

Mauricio quickly leapt into action.

'She turned towards me and said take it off. I didn't think about it, I just followed instructions,' he says.

Mauricio grabbed what he described as a 'spherical blob of bright yellow and blue' in his fingertips, flicked it to the sand and dialled triple-0.

The couple then managed to contain the octopus in a bottle with some water as they waited for the ambulance.

The creature had turned brown with the sun and sand, and Paz was hopeful that, maybe, it wasn't what she thought it was.

But when the ambulance arrived, a paramedic flicked the bottle and the octopus once again flashed blue.

This was an emergency.

Call your family

Soon after the ambulance arrived, Paz began to feel a numbness around her mouth and tongue.

'Like when you eat spicy food but without the spicy feeling.'

In the back of the ambulance, she was told to call her family.

'I think that was a big red flag for me,' Paz says. 'I wasn't afraid, I think I had disassociated. I was being very rational about it.'

In hospital, her respiratory strength, blood pressure and heart rate were decreasing, but overall, she was feeling OK.

Being accompanied by the octopus, nicknamed Cuddles by hospital staff, made Paz something of a fascination during her stay.

'So many people came to visit and take selfies with the octopus. I think that was the weirdest part.'

But it isn't surprising.

The blue-ringed octopus is one of the many dangerous animals Australians grow up fearing. Thankfully, however, very few of us ever interact with one.

When Paz and Mauricio first told their families of their plans to move to Australia, their loved ones were 'terrified' for them, over potential encounters with spiders and sharks.

Paz reassured them, saying 'no worries, I am not going to Queensland' – a place she thought of as home to a disproportionate number of the country's dangerous animals.

Paz's apartment is decorated with photos of her travels around the world with Mauricio, and their cat Uli, who continues to be a great comfort to her following the octopus encounter.

Finding herself belly to beak with one of Australia's most infamous critters still feels hard to believe, but she says it was what happened after the accident that surprised her the most.

In the aftermath, she says two things surprised her: the media's response, and the post-traumatic stress she now experiences, which has prevented her from making a swift return to the ocean, and left her with a disgust for soft, fishy textures.

'I was a big fan of ceviche, so when I visited Chile a couple of months ago my family were waiting with a big pot.'

But Paz couldn't indulge.

News of deadly crocodile or shark encounters, Irukandji stings or in this case, blue-ringed octopus bites, tends to travel very fast in Australia, even making international headlines. It's not uncommon for bite victims to read articles or conversations online centred around their accident.

Paz says she was taken aback by what she saw.

'I felt that it was kind of racist,' she says, adding that people were commenting on whether the victim was a 'foreigner' who didn't respect wildlife.

For Mauricio, this couldn't be further from the truth.

'Paz is the sort of person who can look at the most hideous animal and find in those eyes a spark,' he says. 'She sees our cat Uli's eyes in every other animal. It's alien to me and something that I admire about her.'

Bite Club

Dave Pearson understands this phenomenon all too well, which is why he started Beyond the Bite, also known as Bite Club, a place where those who have had traumatic encounters with animals can share their experiences.

Dave was bitten by a shark in 2011.

Despite the obvious differences between the two encounters, Dave shares a lot of Paz's experiences.

'I started looking at the news stories about myself, and that's when I discovered the not-so-social side of social media,' he says. 'I kind of expected, you know, everyone to go, sorry to hear this happened Dave ... I didn't expect any of the victim blaming.'

Happy to have escaped from the attack with his life, Dave didn't expect to feel anything other than luck when he left the hospital. Instead, he began waking up screaming.

A carefree return to the ocean was also not on the cards. Dave says the stress didn't really hit until six months after he was bitten, when he'd made a physical recovery.

He felt lonely in his experience.

'Nobody has the answers and that was scary,' Dave says. 'Mentally you don't know what to expect.

'A counsellor came and had a few words to me, and I didn't really understand much of what they were saying or take anything away from it.'

In 2018, the Bite Club teamed up with the University of Sydney to research the direct and indirect psychological impacts of shark-bite events.

The study, a first of its kind, found one third of the members of Bite Club who had been bitten by a shark were experiencing post-traumatic stress disorder.

One particular focus of the research was the impact of media exposure on the victims.

The group's experiences ultimately led the researchers to advocate for guidelines for the media, akin to those used for reporting on suicide.

Once upon a time, human interactions with dangerous animals were not so uncommon. And as these interactions have reduced, interest in them has increased, according to lead researcher

Jennifer Taylor, a postdoctoral research associate in the Faculty of Medicine and Health at the University of Sydney.

'People spoke a little bit about not only social media, not only traditional media, but every social interaction,' Dr Taylor says. 'People knew the story, would ask them to retell the story ... and it might have been a space or a place where they actually did not want to get back into retelling their traumatic event. The other side of it too, and we certainly saw this in the shark space ... in a conservation sense, [is] that these topics can get quite political as well.'

In Paz's case, people were quick to point out the dangers of picking up shells, which were potential hiding places for the deadly octopus.

'Unsympathetic and unsupportive conversations can happen when someone's actually healing from a trauma,' Dr Taylor says. 'Certainly, these are valuable conversations to have, but perhaps not with someone who's just survived a traumatic event.'

Dr Taylor acknowledges that the 24-hour news cycle – always hungry for a story – is a challenge, but says journalists must strike a 'delicate balance' between 'the public's need to know and the person's right to heal in private'.

Owning the story

As Dave sought to recover from the psychological impacts of being bitten by a shark, he found one thing made a big difference: talking to other people in Bite Club who had also had traumatic encounters with dangerous animals, and not just sharks.

The group quickly went global with victims of bear, dog, lion and hippo attacks contributing to the conversation.

'I just didn't want anyone to feel alone like I did. And it seems to work really well,' Dave says.

When Dave would meet fellow bite victims, they'd be finishing each other's sentences.

In some instances, Dr Taylor says, this connection to people with similar experiences and 'owning their own story' can be a catalyst for post-traumatic growth.

'Once they'd sort of hit that stage or distance or resolved within themselves that it wasn't the worst thing that had ever happened, it was sort of like ... If I can help someone going through something similar or help someone ease their discomfort, then I'm prepared to do so.'

Paz says she was thankful that when news of her octopus bite went around the world, her name was left out of the stories. But now, she has a desire to share what she's learnt.

Twisting her earring, which is the shape of an octopus tentacle, Paz says, 'If people want me to be the octopus lady, the octopus lady I will be.'

When Paz describes her experience, she goes between the words 'weird' and 'random', but her message is clear. No matter how strange the event, if you know you're not feeling like yourself anymore, it's important to seek help.

This is something Dr Taylor reinforces.

'It's not an aspect of cognition. It's not an intellectual exercise. It's an emotional one,' she says. 'It's about having felt that depth of fear. Even if you did not lose your life, you thought perhaps you could lose your life, and it is that level of fear that is instrumental ... that is what changes biochemistry, the brain, neurotransmitters, all of that stuff. So it's really easy to underestimate yourself and the impact of those sorts of events.'

Determined to return to the water, Paz took 'baby steps' to make that happen.

And last Sunday, she took her first dip since being bitten.

'I'm afraid of this happening again, and I know the probability is very low ... that's an irrational consequence of the accident,' Paz says.

But the allure of the ocean had become irresistible.

'I cried immediately after ... It was like recovering something that I lost, but that has always stayed with me. Something deeply mine.

'I missed it very much.'

✻   *Is this actually PTSD? Clinicians divided over redefining*
         borderline personality disorder, p. **63**
    Why some venomous snakes can bite and kill even when they're
         dead and decapitated, p. **270**

SOME PSYCHEDELIC MEDICINE DEVELOPERS WANT TO DITCH THE THERAPY ASPECT. WHAT COULD GO WRONG?

Rich Haridy

Back in 2016, Rosalind Watts volunteered to work on one of Imperial College London's early trials investigating psilocybin for depression. She had previously spent several years as a clinical psychologist with the UK's National Health Service and quickly saw the incredible potential for psychedelics to help people with mental health problems. By the following year, Watts had abandoned her conventional healthcare work, joining the Imperial team and evangelising the benefits of these miraculous medicines.

Psilocybin is one of several psychedelic drugs currently being explored as treatments for a variety of mental health issues. These drugs, which also include LSD, DMT and adjacent compounds like MDMA and ketamine, can give users profound shifts in perception, both during a dose session and in the weeks or months following. This particular consciousness-altering aspect to the drugs means patients can require more therapeutic support than they would if they were receiving conventional psychiatric medications.

As time passed, Watts grew concerned that these drugs were not being administered with enough broad support for patients. She was finding many clinical trial participants, months later, struggling with the return of their previous sense of disconnectedness and depression, alongside big questions that had been raised in the psychedelic session, with no one to talk to about them. Even with bespoke and careful therapeutic containers created for the trial, the support ended very quickly afterwards.

So Watts and a colleague set up the UK's first community-based psychedelic integration group to provide a safety net to catch any participants in their next trial who might find they needed help afterwards.

'We went and sat in a community centre,' Watts recalls in an interview with *Salon*. 'We gave our time for free, month after month, because many people were having psychedelic experiences without enough support, after trials and retreats, and most months we witnessed people really struggling to cope alone with the aftermath of intense and confusing experiences.'

Disillusioned with corporate impacts on the psychedelic space, Watts changed path to develop a long-term community integration framework for building connectedness to self, community and nature after a psychedelic experience. This was designed to catch people who might struggle to integrate experiences with the drugs after consuming the psychedelic renaissance hype.

In a frank *Medium* post published in 2022, Watts expressed remorse at contributing to a 'simplistic and dangerous' narrative around psychedelic medicine. She still believes psyche-delics are incredible agents for catalysing healing but is also concerned there's a growing narrative suggesting we need to just take the drug and everything will be fine.

Watts wrote, 'The greatest threat to a healthy psychedelic future is the fetishising of just the drug alone,' adding, 'Whether plant, or synthesised compound of one, there is a narrative that all you need to do to change your mind is eat something. I unknowingly contributed to that narrative.'

Don't call it therapy

For the past 70 years, most western medical uses of psychedelics have centered on the way these drugs can serve as an adjunct to psychological therapy, or psychotherapy. From treating alcoholism to alleviating depression and anxiety, the general idea has been that psychedelics can amplify suggestibility, increase neuroplasticity and accelerate the clinical process of psychotherapy.

In fact, for much of the current so-called 'psychedelic renaissance', the treatment has been explicitly referred to as psychedelic-assisted psychotherapy. While there may be more and more research illustrating exactly what psychedelics do pharmacologically to our brain, for the most part it was never particularly controversial to suggest the drugs should always be delivered in tandem with a broad program of psychological preparation and integration.

It's pretty unusual in psychiatry for a drug's efficacy to be fundamentally influenced by counselling or psychotherapy. Certainly medicines such as antidepressants or drugs like buprenorphine for opioid use disorder are known to be more effective when accompanied by clinical support – but they are often administered solely as pharmacological treatments.

Psychedelics, on the other hand, are considered to be so life-altering, emotional and sometimes distressing that having a licensed practitioner nearby to walk one through the experience has been one approach to reducing harm or side effects.

Over the last few years, however, the word 'therapy' has been appearing less and less in psychedelic clinical trials as the research moves closer to real-world applications and approval by agencies like the US Food and Drug Administration (FDA). Setting aside the somewhat atypical example of MDMA-assisted therapy for PTSD – a treatment modality that intrinsically intertwines psychological interventions into its process – recent clinical research with classical psychedelics (psilocybin, LSD and DMT) has quietly dropped the psychotherapy and replaced it with a variety of more ambiguous terms such as 'psychosocial support'.

Compass Pathways has led the way in this semantic shift. Currently deep in Phase 3 trials testing psilocybin for depression, the company has been forceful in its reframing of psychedelic medicine away from the classic psychedelic psychotherapy model. Compass studiously avoids any reference to psychotherapy in its trial work, instead calling its research 'psilocybin treatment with psychological support'.

A provocative 2022 commentary published in the *American Journal of Psychiatry* and co-authored by Guy Goodwin (Compass's chief medical officer) and Ekaterina Malievskaia (Compass's co-founder) made the case for psilocybin treatment being a drug therapy first and foremost. They argued support to ensure psychological and physical safety was of course necessary with psilocybin treatment, but their dose-response data purportedly indicated efficacy was primarily a pharmacological effect and more expansive psychotherapy programs were not necessary. In other words, the drug does most of the work.

Part of this subtle shift in nomenclature has been informed by the FDA's first guidance for psychedelic clinical studies, released in mid-2023. The guidance established an assortment of clinical issues the drug regulator will be considering when looking at applications to bring psychedelic medicines to market.

One crucial question the FDA raised in the document was what it saw as a problematic conflation of drug effect and psychotherapy when exploring the therapeutic efficacy of psychedelics. It suggested clinical research should attempt to separately quantify the contribution of psychotherapy from the drug's effects when evaluating a patient's response to the treatment.

Essentially the question being asked was: are these drugs still safe and effective in the absence of psychotherapy?

A recent advisory committee to the FDA was particularly focused on this question when trying to evaluate the safety and efficacy of MDMA for PTSD. This problem with psychotherapy ultimately played a part in the committee's near unanimous recommendation against the approval of MDMA therapy, which could dampen hopes that the drug will be legally prescribed in the near future.

The drug does all the work?

In early March 2024, psychedelic startup MindMed made a pair of striking announcements. It revealed the FDA had granted breakthrough status designation to its investigational drug MM120, a patented formulation of LSD. It also revealed data from its Phase 2b study testing MM120 for generalised anxiety disorder. According to the company, a single dose of the psychedelic drug led to significant and sustained clinical improvements for up to 12 weeks.

Less prominent in MindMed's announcement, but perhaps most significant, were details surrounding the treatment protocol being tested. MindMed was claiming to have completely eliminated all traces of psychotherapy from its treatment model.

In an investor presentation the company proudly declared its MM120 trial contained no 'preparation', no 'integration', and no 'ongoing therapeutic engagement'. This was psychedelic medicine pared back to the absolute minimum. Spend a day taking the drug in a safe clinical environment and then go back to your life.

Speaking to *Salon*, MindMed's chief medical officer Dan Karlin said the primary goal of the trial was to home in on the sole pharmacological effect of the drug, as suggested by the FDA. Every step of the trial design looked to remove conventional traces of psychotherapy while maintaining basic levels of support necessary for patient safety.

'There are elements of the history of these drugs, like the need for psychotherapy, that we were doing our best to try and test if that is a real need,' Karlin explained to *Salon*. 'Not because we don't think people would benefit from it, because psychotherapy helps. But because we wanted to see if the drug worked without it.'

So instead of a series of preparatory therapy or support sessions before a drug dose, MindMed simply engaged with the patient in an expanded informed consent process. According to Karlin, informed consent for psychedelics looks a lot like it would for any other psychotropic medication except it is maybe 'a little more robust'.

'The conversation there is about what people may or may not experience, from a perceptual standpoint, from a cognitive standpoint, from a physical sensation standpoint, and from an emotional standpoint. Then there's a putative conversation about mechanism,' Karlin said, describing MindMed's informed consent process.

In the room for a dosing session there are two 'monitors'. One of those monitors is either a medical doctor or a nurse practitioner, while the other need only have a bachelor's degree with some working experience in anything 'adjacent to psychiatry'. Some of the session monitors used in the MindMed trials did have experience with psychotherapy. However, that was not a requirement when selecting personnel. In fact, Karlin explains some of those monitors in the trial were even medical doctors from non-psychiatric fields such as oncologists and gastroenterologists.

'In the room, these folks are instructed explicitly not to provide psychotherapy,' Karlin notes. 'What they do provide is monitoring obviously for safety, though that has not really been much of an issue. They do things like vital signs, [other health] checks and then they assist the participant if they need anything.'

Of course, sitting in a room with someone going through a high dose psychedelic experience is rarely a completely passive experience. Karlin admits there were moments where participants were in distress or looked for some kind of engagement with the monitors. In those instances, the monitors were instructed to reorient the participants to what's happening in the room and remind them the drug's effects will pass.

'In essence, it's supervising, not therapy,' he stresses. 'Most patients on very high doses are not looking for dyadic engagement. They're not looking to be psychotherapised. Occasionally, someone would say something or ask a question that might be sort of leading towards something that could be a more therapeutic intervention. And then the instruction that the DSM [dose session monitor] had been given was to redirect back to the patient, they have to do their thing while they're here.'

Once the dose day was over, the participants went home and were followed for several months to track the effectiveness of the treatment. Two in-person assessments were conducted in the week after the dose. Here, the participant met up with one of the monitors present during their dose. Karlin stresses no 'systematic psychotherapy' took place at these follow-up appointments. The participants simply completed assessments to measure their symptoms and occasionally asked the monitors about events that may or may not have occurred during their dose session.

So has MindMed demonstrated a psychedelic drug can be clinically effective in the absence of psychotherapy or a broad psychological support protocol when treating certain psychiatric conditions? Does this mean the future of psychedelic medicine is a clinic where patients drop in, spend a day experiencing a drug while under supervision, and then simply leave with no follow-up support or management necessary? Maybe.

But what if this therapy versus drug binary is the wrong way to think about the impact of psychedelics? Is this dynamic a red herring that could be diverting our attention away from other, more deeply metaphysical issues that can emerge with psychedelic use? If researchers are only measuring a drug's efficacy using anxiety or depression scales and tracking adverse effects from the perspective of things like suicidality, could they be missing other kinds of long-term adverse impacts?

Ontological shock

'My sense of self was gone. It was like it had been obliterated into a million pieces,' Theo said. 'And I couldn't really work out who I was. I just had this, like, routine of going to work. And I just stuck to that as a means of like, keeping some sense of inner structure by having some outer structure. And, like, if I'm honest, if I'd sat and thought about that for a while, for the first few weeks, I think I would have just cried because like, it didn't seem very clear who I was anymore.'

Theo participated in an ayahuasca ceremony in his mid-20s. His experiences in the weeks and months following the psychedelic brew led him to profoundly question his identity. Theo is one of 26 people interviewed in a new preprint study exploring the varieties of existential distress that can follow a psychedelic experience. There's Adriene, a woman in her thirties who, after a magic mushroom experience alone in her apartment during the pandemic, changed her life path from being an atheist dominatrix to taking vows and becoming a Buddhist nun. There's Cal, whose experience with DMT and cannabis at a party moved them away from a journey to become a rabbi and into a world of Wicca and paganism.

Western science has a name for these radical transformations, coined in the early 1970s by philosopher Paul Tillich: ontological shock. Tillich's work was focused on how we as humans conceive of God and what happens to us if we are confronted with God's non-existence. One of his definitive works concluded with the iconic line, 'The courage to be is rooted in the God who appears when God has disappeared in the anxiety of doubt'.

Tillich suggested certain experiences in life can trigger a kind of spiritual destabilisation where people are confronted with existential realisations. This ontological shock is precipitated by challenges to one's worldview or beliefs. For Tillich, an ontological shock was primarily contained within a theistic frame. So something like a near-death experience, for example, could bring a person face to face with a feeling of there being no God in the universe. This profound break in a person's belief system can be frightening for many, leaving one feeling unmoored, disconnected and groundless. Over the last few decades, the term ontological shock has been more broadly used to describe experiences where someone's fundamental metaphysical beliefs are deeply challenged beyond theism. Members of alien abduction communities, for example, often speak of the ontological shock they experience after an abduction encounter.

Psychedelic experiences are also deep mediators of ontological shock. From a clinical perspective, these kinds of psychedelic-induced ontological shocks are not merely a bothersome side effect of the drug, but instead could be crucial to the kinds of healing being seen in medical contexts. 'The therapeutic benefits of psychedelics are theorised to be driven by increases in entropy – a measure of uncertainty – that shift individuals' reliance on their prior beliefs,' the researchers write in the study. 'This destabilizing mechanism allows for a recalibration of cognitive structures, enabling the re-evaluation of previously rigid mental models. However, when individuals lack adequate psychological or social resources, this increase in uncertainty can lead to distress manifesting as confusion and difficulty accommodating the ungrounding of established worldviews.'

Psychedelics in a disconnected world

Most of the work Ros Watts does nowadays is in establishing systems of support for people after psychedelic experiences. She co-runs an online integration community called ACER Integration that offers peer support and a 12-step process for building connectedness. The big problem, from Watts's perspective, is that 21st-century western culture is disconnected and isolating, lacking any kind of context or container to help people safely integrate psychedelic experiences.

'If you put psychedelics into a disconnected culture, then people go back to disconnectedness, and they're left with these big huge questions with no one to talk to,' Watts says.

A reigning aphorism of the psychedelic renaissance is that these drugs have been safely used by innumerable indigenous cultures for thousands of years. And while that may be true, Watts is keen to stress a big difference between every psychedelic-using community in the past and people in the world today.

'Indigenous groups have [used psychedelics] as part of a community,' Watts explains. 'You have loads of support, you're doing it in a group of people, you have your shaman, you have your community care. There's no "psychedelic integration", because you're living in a community of people who are integrating all the time. After a psychedelic session people need a community of ongoing support from people who understand. Not just to mitigate risks, but to maintain benefits. We can't extract the medicine from its context of doing it that way.'

From MindMed's perspective, Karlin doesn't deny some people can have challenging experiences with psychedelics, but he stresses drug developers do not control the practice of medicine. And neither does the FDA, he adds. By establishing a potential baseline of safety and efficacy in their clinical trials, Karlin argues MindMed is helping broaden access to these medicines if they were to be eventually approved.

'The goal is that in the real world this is just another tool in the toolbox ... to bring this toward regulatory approval in such a way that a label and REMS [Risk Evaluation and Mitigation Strategy], if one is required, are as much as is needed for true safety, but broad enough that this can be incorporated into a lot of different types of practice patterns.'

Karlin agrees psychedelic drugs can lead to patients making large life changes due to the nature of the experience. But he doesn't see this as something wholly novel to psychedelics. He argues patients undergoing psychotherapy, for example, can often come to similar major life-changing realisations.

'People getting good psychotherapy sometimes leave their families,' he says. 'That's the nature of exploring one's own mind. These drugs can accelerate that process in many cases, but they don't change someone. They just open people to the idea that maybe they're not on the path that's right for them now.'

Watts is a little more concerned about the shockwaves likely to ripple through communities if these drugs are simply slotted into a western medical model without broader support systems in place. Her experiences with people who have been deeply unsettled by psychedelic trips led her to believe small, grassroots organisations are going to have to pick up the pieces when big pharmaceutical companies start rolling these medicines out. And she feels pharmaceutical companies making profits off these drugs have a larger responsibility to at the very least offer financial support to these organisations.

'The list of deep, powerful experiences where you need a hand to hold and need someone to work with you afterwards is very long,' Watts says. 'So what [these companies] could do if they had compassion is support people through those processes. And not gaslight them by saying, "Yeah, this [single drug dose] is going to fix you." What they would also do is they would provide a proportion of their profitss to support grassroots organisations who have been set up to catch the people who are damaged by psychedelic therapy.'

✱ *A freediver finds belonging without breath*, p. **69**
Insect consciousness, p. **87**

A SNOWFLAKE IN HELL

Jackson Ryan

'YOU ARE SIMPLY A FUCKING WEAK BRAINWASHED MILLENIAL.'

That was how I met John.* On 27 November 2019, John sent an email to my work address titled 'Message from a reader', the default subject line for an automated message sent via the web. John was writing in response to a report I had written a day earlier regarding a 2019 study by the World Meteorological Organization. The WMO, eminent in the study of climate and weather, had determined atmospheric carbon dioxide levels in 2018 were the highest recorded in human history. This was not a surprising find for scientists. It was expected. In the years since that report was published, the levels have only risen higher.

The basic science of climate change is not up for debate. The unfortunate reality is that greenhouse gases emitted from the burning of fossil fuels, such as carbon dioxide, cause the planet to heat up. As a result of the excess CO_2, the average temperature on Earth is about 1.2 degrees Celsius higher than it was during the pre-industrial era, around 1850. Those temperatures continue to rise – 2023 was the warmest year on record.

Back in 2019, the report angered John.

'WE NEED CARBON DIOXIDE FOR PLANT LIFE YOU IDIOT.'

---

*   Names in this essay have been changed.

This type of abuse is the response du jour of climate-change denialism. It's a vitriol encountered almost daily; whenever you write about the climate, carbon dioxide levels, reports from the Intergovernmental Panel on Climate Change – the list goes on – you must buttress yourself against conspiracy, pseudoscience and abuse. In other words, John's response wasn't special. It wasn't different. It was an online Groundhog Day.

Typically, I'd ignore the email and simply click delete but outside my office window, ten storeys above a Starbucks in Sydney, the air was a muddy brown. Bushfires at the fringe of the city, burning out of control, had delivered a thick blanket of muck over the streets below.

The horizon took on the colour of a menacing sandstorm. Sydney Harbour, usually visible from my office desk but now enveloped in a fog of ash, was haunting. During one lunch break, I walked to the Opera House and I could taste it in the air. I saw runners circle the iconic white sails of the venue, soaked with sweat. I thought about their lungs – my lungs – choked with soot. And I thought about John.

John, I'd soon learn, lived on the other side of the Pacific Ocean. He wasn't experiencing the minute-to-minute effects of a smoke-clogged city. He wasn't tightening the strings of an N95 mask as he lined up for coffee, or checking the Air Quality Index on his phone every hour. He was sitting at home in Dallas, Texas, while he bashed out abuse on his keyboard or phone.

I didn't delete the email. I took a screenshot of it, erased the identifying information and posted it to Twitter. A follower on Twitter urged me to respond. That was all the convincing I needed.

'OK boomer,' I wrote to John, and hit send.

About two weeks later, John replied. He was not pleased. The second email was more aggressive than the first, calling me a 'SNOWFLAKE' and explaining that he owned four businesses (automotive), routinely had intercourse with his wife (I'm paraphrasing), and how, if he was my father, he would 'slap the shit out of me'. I posted this response to Twitter. It was funny. Someone even took my tweet and posted it to Reddit, where it became the second most popular submission in the tongue-firmly-in-cheek 'I Am Very Badass' subreddit.

My jousting with John continued late into 2019, covering a wide variety of climate topics. We did not see eye to eye on … well, anything. I had fallen deep down his rabbit hole, arguing against erroneous and belaboured talking points of climate-change denialism that have existed for decades: the idea CO_2 was inconsequential, the definition of ice ages, the normality of bushfires, and why some parts of the world were getting colder. But as we corresponded towards the end of 2019, John softened.

I began to soften, too. I wasn't interested in personal jabs. I could tell John was far more reasonable and thoughtful than his initial email suggested. I really wanted to find some common ground, to try to understand where his beliefs came from.

Two days before New Year's Eve, my inbox dinged again. John didn't believe a lot of what I was telling him, but he conceded that he couldn't 'know everything' and said he would like to know more about the climate. He asked if I could help in some way. I was more than happy to provide him with the facts, as science had produced them and as I understood them. What started as combative, sarcastic and frustrated became charitable, understanding and empathetic.

Then COVID-19 arrived. Climate change became a secondary issue. During the pandemic, John and I would sometimes correspond about the virus or lockdowns. He would link to obscure websites about the dangers of COVID vaccines or discussion about debunked treatments such as colloidal silver. I'd try to refute some of the wilder claims and keep looking for common ground. I'd even come to learn his wife had received two liver transplants – she was immunocompromised – but that John refused to wear a mask or get the vaccines.

I struggled to make sense of John's contradictions and beliefs.

I wanted to understand how he'd come to rally against climate change and vaccines; I wanted to know why the strongest scientific evidence meant little to him and why he was so distrustful of the experts.

So, almost three years to the day since he first emailed, after flying to the United States, I took an Uber to a Tex-Mex restaurant in uptown Dallas, ordered a tequila-based 'Mambo Taxi' and found a spot on the upper deck, outside in the sun. I waited, wearing a knitted sweater with the phrase 'SWEATER ENTHUSIASTS FOR CLIMATE CHANGE' stitched on the back, and dipped corn chips into salsa, a little nervously.

John had agreed to meet. He was just a few minutes away.

It's 17 minutes to midnight in 1991. The symbolic hands of the Doomsday Clock had been moved back seven minutes to 11.43 pm following the collapse of the Soviet Union and the end of the Cold War. The timepiece, maintained since 1947 by the *Bulletin of the Atomic Scientists* and residing at the University of Chicago, is intended to symbolise humanity's proximity to global catastrophe.

It was never designed to be further than 15 minutes from midnight, but as the *Bulletin*'s scientists wound its hand back in 1991, they proclaimed the world had entered 'a new era'. The threat of global conflict was receding at the Cold War's end.

They were right. It was a new era, but their optimism was off the mark. Though global warming hadn't regularly made headlines in 1991, the planet was heating up at a much faster rate than we would be able to handle. The burning of fossil fuels, such as coal and oil, generates huge amounts of carbon dioxide. The molecule is simple: one carbon atom, two oxygen atoms. A molecule of CO_2 is a gas at room temperature, bumping around and through the atmosphere, where it can reside for decades to hundreds of years. A quirk of its structure allows visible light from the sun to pass through it, but infrared light gets absorbed, releasing a chunk of energy, as heat, back towards the Earth.

Our planetary heating system has been known since the 1820s, when French scientist Joseph Fourier suggested sunlight alone would result in a much cooler Earth than we experience. Scientists have shown that without CO_2 the planet's average temperature would likely be colder than −18 degrees Celsius, and there's not much life (at least as we know it) that could thrive in those conditions.

It's sometimes been termed the 'greenhouse effect', and carbon dioxide, along with methane, water vapour and nitrous oxide, are termed 'greenhouse gases'. Together, the gases have ensured Earth's average temperature sits at a much cosier 15 degrees Celsius.

But, since the mid-1800s, levels of those gases were no longer entirely controlled by the planet's natural cycles. Thanks to the industrial revolution and the rise of agricultural production, humans began pumping those gases into the atmosphere at an increasing rate.

It wasn't until 2007 that the words 'climate change' first featured in the Doomsday Clock's annual proclamations of potential apocalypse. The same year, Al Gore and the United Nations Intergovernmental Panel on Climate Change received the Nobel Peace Prize for 'their efforts to build up and disseminate greater knowledge about man-made climate change', including Gore's documentary film, *An Inconvenient Truth*.

But climate scientists had been sounding the alarm long before the Doomsday Clock was wound back in 1991. They were concerned with the extreme rise in atmospheric CO_2 and other greenhouse gases for decades prior. Even fossil fuel companies, such as Exxon, had commissioned studies that demonstrated how its business could be catastrophic for a substantial fraction of life on Earth in the future.

By 2007, the IPCC concluded Earth was already feeling the

effects of a changing climate. Average temperatures had risen by about three-quarters of a degree. Life at the poles was impacted by differences in sea ice coverage, some migration patterns were being noticeably disrupted, glaciers were melting at unexpectedly high rates and biodiversity was suffering. This was exactly as some earlier models had predicted.

Climate change had already begun to alter the planet in ways scientists expected – as well as ways that were completely unexpected.

John was also just as I expected and completely unexpected. Tall, shoulders as wide as the door frame he walked through. Slick wet hair, slightly curled. He shook hands firmly, as if he was juicing a lemon. If you were to catch a glimpse of him from across the room, you might mistake him for American entrepreneur Mark Cuban – if Mark Cuban had wrestled for his entire adult life.

John didn't come alone. He also brought his wife, Jane*, who exuded a Southern charm that was disarming and kind; a calm wind next to her hurricane husband. When the pair sat down at my table, John hollered at the waiter for queso and dip, extolled the virtues of Tex-Mex cuisine and explained the best way to do nachos, while discussing the latest news – at the time, it was the attack on the husband of former Speaker of the House in the United States, Nancy Pelosi.

'So you're not going to slap the shit out of me?' I asked, with a smirk.

We were well past that point. John is a climate-change denier. To say we didn't see eye to eye on climate change would be understating it. But our email exchanges had become friendly jousts. In person, he was reasonable, buoyant and listened to what I had to say without interrupting. He made me laugh. He was confident (perhaps to a fault). Did I … like this man?

His political views were scattershot. He called Trump an asshole, but believed the former president had done wonders for the US economy. He hated Congress – both sides of it – and the divisiveness that prevented consensus. His disillusionment is typical of the American experience in the last decade. Trust in the government is at historic lows, with only about one in five Americans saying they trust Washington to do what is right 'just about always' or 'most of the time', according to the Pew Research Center. In Australia, that number is about two in five.

John's distrust of the US government manifested in various ways. It wasn't only climate change. He proudly confirmed he hadn't been vaccinated against COVID, and he told a story about the time he visited a supermarket, boasting a lanyard around his neck that signalled his right to remain unmasked to any intrepid staff member who might question his act of civil disobedience. He was also increasingly concerned about immigration into Texas from across the US–Mexico border.

He could get on a roll, too, moving from one twisted factoid to a hardened truth and through streams of questionable science. Sometimes, Jane would agree; other times, she'd roll her eyes and laugh in a 'here he goes again' way. Even within their household they disagreed on facts. Jane was agnostic about climate change.

We talked well into the evening, hopping over to a bar across the street and pulling up three stools. We covered Australian colloquialisms, the meaning of 'mate' (and our friendly use of the word 'cunt'), more politics, more climate change, their kids and both of their careers. I was confident John no longer thought of me as a 'fucking weak, brainwashed millennial'. He might have even called me mate. With a little time and practise, maybe even cunt. But I hadn't changed his mind about climate change.

Two days later, on a warm November afternoon, I departed Dallas for Rapid City, South Dakota. At the airport, I received a text from John. 'Have fun in SD,' he wrote, 'it is beautiful country.'

The Badlands of South Dakota provide one of the most uniquely beautiful backdrops in the world. The sweeping prairie is crowded with spires of rock jutting like gnarled fangs into the sky. Driving through the area's winding roads, between the buttes and towers of the ancient landscape, I see prairie dogs standing at attention from mounds in the earth and bighorn sheep calmly chewing on grass as they follow the path of tourist cars.

The Badlands National Park, which covers almost 100 000 hectares of protected country, was once home to a vast inland sea. The water provided it with its defining characteristics. Sediments began to build up, layer by layer, during the time of the dinosaurs some 75 million years ago. As the water drained away, deposition of these sediments slowed. Around 23 million years ago, it stopped altogether. The remnants of that era are tattooed into the rock as thick bands of colour, ochre lines reaching around every natural formation. Over the last 500 000 years, the crooked teeth of the Badlands' spires have gradually eroded.

As the rock whittles away, monsters emerge.

The region is a treasure trove of fossils, a vast graveyard of extinct mammals. Remnants of mammalian precursors such as creodonts, the pig-like entelodonts, camels, rhinos and even the fearsome mosasaur, which stalked the sea during the final age of the dinosaurs, have been discovered in the Badlands. These fossils were first recognised by the Lakota people, a group of seven Native American tribes that have lived off the prairie, hunting in the area for upward of 11 000 years. 'Bad lands' is translated from the Lakota term mako sica, a sobriquet borne of the region's untraversable and shifting surfaces, its deep valleys and unscalable cliffs.

It's also a somewhat unfortunate name because John is right: it is beautiful country. And climate change threatens that beauty. An analysis by the US Geological Survey and the National Park Service in 2016 examined four different climate futures in an attempt to plan how the Badlands' vast resources will be managed as the planet warms. Some of the outcomes present grand challenges for the park's most famous residents, the American bison (often incorrectly dubbed 'buffalo'), which are able to roam freely across the Great Plains.

Bison – tatanka or pte – hold a special place in Lakota culture. Nomadic tribes found everything they needed in this one animal: its meat was used for sustenance, its shaggy winter coat supplied raw material for clothing, and its bones and horns, whittled down, provided tools or toys. The bison were not just resources; they didn't just sustain life. They gave life. They *were* life.

As the temperature rises, it's predicted there will be less space for bison to forage and find water. This could bring herds closer together, increasing the risk of disease jumping from animal to animal. Other scenarios predict increased rainfall, which could see an explosion in tick and mosquito numbers, with similar consequences for bison health.

The monsters in the rock will suffer, too. It's expected that warmer temperatures will promote extra weathering and less vegetation cover. As the bones of ancient creatures slowly emerge, they're more prone to looting and disturbances. The Badlands is a library of deep time, and the femurs, skulls and spines of long extinct mammals are the books that reveal its secrets. Climate change in the Badlands doesn't just result in the degradation of John's 'beautiful country'. It means erasing cultural ties to the land, the creatures it sustains and the ones preserved within its rocky walls. It means erasing history.

With the sun's fingers receding from the highway out of the Badlands, I decide to make a left turn onto a rocky dirt road that my Toyota Camry is ill-suited for. A ranger had mentioned the park's bison often roam the area. I hadn't seen any during my day scoping the buttes and crevices, walking to the very end of dangerous trails where signs usher you back the way you came. But I figured I'd give it one last try.

After driving for about 15 minutes, with darkness fully settled over the prairie, I noticed a long shadow stretching across a vast, empty plain. A slow-moving patch of darkness with horns. A herd. There must have been two dozen or so bison; bulls, cows and calves padding carefully along the short grass with no urgency under the bright disc of the moon. I pulled over, shut the engine off, but left the headlights on. I waited.

The herd eventually blocked the road ahead, a few metres from my Toyota's grille. A bull calmly turned its head towards the car. In the moonlight, it simply stared, its eyes reflecting red in the car's lights.

The stark red soil of the Australian desert might be worthy of a souvenir shop's postcard display, but it's the country's lesser-known heathlands and forests that present a more subtle beauty. Only half a year after the devastating fires were extinguished in 2020, a friend books us into a privately owned campsite just past Blackheath, on top of the Blue Mountains that run down Australia's eastern flank.

When we arrive, George, the owner, comes out to meet us in his driveway as if we're long-lost friends. He's jovial and friendly with a stereotypical Australian twang; a sprinkling of Crocodile Dundee-isms in his speech. There's a little Saint Nick about him, too, though he lacks the lengthy, white beard. Without prompting, he begins to describe his property, explaining how it stretches across a gaping gully, rife with eucalypts and prickly paperbark.

Driving down to the campsite, I'd taken note of the landscape. The trees are painted black from the neck down; charred by fire. The same fires that, at the end of 2019, had draped Sydney in ochre smoke. On this day, the sky is dark, ready to burst. Before setting up, we need to go get ice from the local pub and, on the way out, I catch George. I ask about the fires. Would George, too, think I'm a weak, brainwashed millennial?

After a brief exchange about arson, I cut to the chase. 'Do you believe in climate change?' George shakes his head, rubs his chin. 'The climate has always changed,' he says. 'Nothing we can do to change it.' I don't respond, I just nod along. I don't want to get the owner of the property offside. He's shown nothing but kindness. As it rains, he even offers us a spot inside until the storm passes. And if my experience with John taught me anything, it's difficult to change someone's mind about climate change, even armed with facts.

Research by the University of Queensland's Matthew Hornsey and Kelly Fielding shows that to be true. In a 2016 meta-analysis, published in the journal *Nature Climate Change*, the duo and two other researchers synthesised data from 25 polls and 171 academic studies internationally and found – perhaps, unsurprisingly – that 'ideologies, worldviews and political orientation' were associated with belief or rejection of climate change.

In a Zoom interview with Hornsey, a professor of social psychology, he shows me a model he created to try to understand why people come to reject science. The model takes the shape of a tree, with branches haphazardly reaching for the sky and spindly roots twisting down into the earth. Those branches represent the surface attitudes towards science – that climate change isn't real or genetically modified food is wrong. But, to find out what makes these attitudes resilient in the face of contradictory evidence, 'you've got to look at what's underneath the surface', Hornsey says.

The root attitudes – the ideologies, worldviews, interests – are deeply seated and tied to a host of personal and social identities. In essence, Hornsey's model suggests that to change George's mind, I'd have to change his entire worldview and appeal to a value system different to my own. Presenting the facts alone would not suffice. 'It's not that they haven't heard the data,' Hornsey says. 'It's just, for whatever reason, they don't want to believe the science.'

George, like John, was affable and friendly. He was accommodating. He wasn't stupid or uneducated. He seemed to like me. If we'd engaged online, perhaps it'd be different. While my friend ducked inside to use the toilet, George and I continued to talk about his property – its location, when it was built, the beautiful country at his doorstep and the highway that lurked just a few feet from where we stood. He genuinely seemed to love the land.

But as a ute flew past his front gate, I watched as George made a gesture with his boot.

'If you want to build a road and there's an endangered frog in the way,' George said, gesturing at his foot, lifting it up, stamping the imaginary amphibian into the wet cement. Squish. I stared at the empty space where George's frog lay, picturing its internal organs messily sprayed across the concrete like the yolk of a smashed egg.

A few weeks before Thanksgiving, I buy a bacon and egg bagel from a bustling Harvard cafe and drive from Boston to Falmouth, Massachusetts. Once I get out of the city, the highways are lined by immaculate green, gold and red; shades of the final days of autumn.

I pull into a gravel driveway on an idyllic street filled with houses that have been standing since the end of the Great War. James Manning is standing out on the wooden deck to greet me. His wife, Robin, smiles as I attempt to park and re-park the car in just the right position, leaving enough room for her to get into the garden.

In October 2022, Manning retired from the National Oceanic and Atmospheric Administration after spending 35 years as an oceanographer, a large proportion of which he spent at NOAA's office in Cape Cod, the peninsula of quaint, historic villages extending into the Gulf of Maine.

Manning tells me his retirement was to spend more time with his grandchildren, one of whom squeaks around the deck with a toy. After sharing a turkey sandwich and discussing the finer points of toddlerhood, Manning and I hop into his van and head to the cape together, towards the esteemed Woods Hole Oceanographic Institution and his old workplace. We soon find ourselves strolling, in whip-cold winds, alongside the Woods Hole aquarium, where a one-eyed rescue seal named Kitt dashes around its enclosure.

Manning's love of the ocean is obvious – he points out historical facts about the region, and guides my eyes towards the most interesting organisms occupying the aquarium and explains the roles they play in the ecosystem. His love for the area started early. During a camping trip around the cape at nine years old, his father, Joseph, woke him in the middle of the night. 'Come with me,' he said. The duo drove up to Provincetown, a village situated on the coast right where Cape Cod's finger-like headland curls back in on itself. His father had arranged a trip with the captain of an old fishing boat, the *Plymouth Bell*, to take the pair out on the ocean. Manning most fondly recalls eating a huge bowl of cornflakes in the galley, but it's clear his time on the *Bell* had a lasting impact on him. These days, though retired, he still works with commercial fishermen to understand the ocean.

Manning's core focus has been on measuring the temperature at the bottom of the Gulf of Maine. With the help of local lobster fishermen, who attach simple electronic recorders to their lobster cages, he has been able to record changes over decades. In collaboration with local schools and universities, he's also helped develop GPS-tagged drifters, often shaped like turtles, which float along the ocean surface to study currents. In short, if anyone could see the effects of climate change in real time, it's Manning. He has watched, slowly, as the red lines on his temperature graphs tick upwards.

'The big story there is that the Gulf of Maine has been warming more so than most places in the world,' he says.

The gulf is fed by the cool waters of the Arctic via the Labrador Current, and the warmer waters of rivers and streams across eastern North America. But the Labrador Current has been weakening and the cooler waters are no longer making their way into the gulf. The gulf's ten hottest years on record have all occurred in the last ten years, with the warmest of all coming in 2021 and the second warmest in 2022, according to the Gulf of Maine Research Institute.

The impacts of warming ripple through the gulf's food chain, disturbing the tiniest species – microscopic drifters known as zooplankton – all the way up to the critically endangered North Atlantic right whales, of which fewer than 400 individuals remain. The lobster fishermen have to deal with changing conditions and, thus, breeding grounds for their catch. Those conditions are already changing, with dire consequences.

In September 2019, some of those fishermen began to pull up lobster traps full of the dead. Not only lobsters, but lifeless fish and crustaceans filled the rusty cages. A team of scientists from Woods Hole and the Massachusetts Division of Marine Fisheries quickly established the culprit: a 'blob' of water, about 9.5 kilometres wide and at a depth below 10 metres, that had been stripped of oxygen.

What had caused the blob? Scientists believed the dissolved oxygen levels were related to a number of factors. Warmer waters in Cape Cod Bay contributed, but so did changing wind conditions. As a result, a harmful type of algae, known as *Karenia mikimotoi*, was able to flourish. The algae has existed at low levels in Cape Cod Bay for decades, but, in 2019, the conditions were right for it to bloom. The algae, the temperature and the wind conspired to strip the 'blob' of dissolved oxygen. The fishery cages became death traps.

Lobsters, fish and crabs could not outrun the blob. Upon the ocean floor, they suffocated.

We are all, in our own way, contradictions. Though it wasn't the sole purpose of my journey from Sydney to Dallas, meeting with John meant taking a 14-hour, carbon dioxide–spewing flight across the Pacific Ocean. I had to get an Uber ride, 40 minutes in total, from my hotel to the Tex-Mex restaurant where we ate. The SUV I rode in was powered by petrol. Then there's the queso, made of cheese, a product of dairy cows, a farm animal responsible for a huge chunk of the greenhouse gas emissions from livestock. And I was wearing a sweater, made of 100 per cent merino wool, which also contributes to humanity's ever-growing carbon footprint. I needed an Uber home, too.

We're all in some ways guilty, then, but there are much bigger influences at play. Since the 1990s, many of the world's biggest fossil fuel companies have waged coordinated and successful misinformation campaigns. The very idea of a 'carbon footprint' – a somewhat flawed way to assess individual impact on the environment – was popularised by a marketing agency hired by British Petroleum in the early 2000s. Subtly, individual action was deemed more important than collective action.

Regardless, there remains, particularly in the United States and Australia, a great divide between the nominal left and right when it comes to belief in climate change. If you identify as a Republican, there's only about a 23 per cent chance you will also be concerned about climate change. It's similar for conservatives in Australia. That divide seems to be widening.

For Hornsey, the University of Queensland researcher, there are opportunities to change attitudes with a technique he calls 'jujitsu persuasion'. In jujitsu, a competitor uses their opponent's momentum and weight against them. An opponent's awkward lunge becomes a graceful roll onto the mat. A defensive stumble becomes an offensive submission. Hornsey suggests a similar tactic might be useful when dealing with conservatives who dispute the facts of climate change.

I could argue the physics and scientific basis with John or George, describe the interaction between carbon dioxide molecules and light, what an atmosphere is, the greenhouse effect and any other climate phenomena I could think of. Even with complete recognition and understanding of the science, they'd argue straight past it. It's not a question of science, but identity. John would sooner see the 'beautiful country' of South Dakota disappear than change his mind on climate change. George's life had been directly affected by fire, but even that couldn't shake his beliefs about the environment. I'd rubbed the soil of the Badlands between my fingers, heard the wind offer secrets to the burnt trees of the Blue Mountains. I could hardly reconcile my experiences of the natural world with theirs. Or could I?

At the restaurant, with John, I started to realise I couldn't change his mind with facts or science. Beating him over the head with the science or the truth or the facts was barely registering. It was a Tex-Mex stand-off and we would continue talking past each other. Wouldn't we?

On 28 March, another email. The subject line? 'The Physics of Climate Change and Proof that CO_2 causes the warming.' This, at first glance, seemed like progress.

After three years, I'd done it. I'd convinced a sceptic, finally, of the science that shows our world is warming. All that time and energy was not for naught.

That was not the case.

John's email pointed to an abstract, posted on the American Meteorological Society website, by a retired US Geological Survey scientist known as Peter L Ward. This is not a name I had encountered previously, but a quick google revealed that Ward has long been promoting his idea that warming is not in fact caused by CO_2, but by ozone depletion. While dozens of studies have shown this depletion does not cause temperatures to rise, the abstract John had found appears quite convincing. It's on an official website dedicated to meteorology, after all. Ward's credentials were sound, but he wasn't studying climate – he was studying volcanoes.

I provided a long-winded response, focused on the claims John raised, explaining the provenance of the information he'd found. The abstract was from a conference, in 2018, at which Ward had presented a poster of his own at-home experiments with styrofoam boxes and a black pot of water, and erroneously claimed 'increases in greenhouse gas concentrations have never been shown by experiment to cause observed global warming'. Those studies *have* been performed, but the cherry-picked conference abstract gave no indication of this.

After three years, I can't help but wonder if the entire approach is wrong: perhaps I should begin incorporating the idea of jujitsu persuasion into my thinking. To change John's mind – if that's even possible – means changing his entire worldview. It means trimming back the roots, not pruning the leaves on every individual branch of his beliefs. In a world where Facebook groups, Twitter/X hot takes and TikTok tantrums have become the go-to methods of delivering information, this is a task that feels insurmountable.

Standing on my bookshelf in a simple, black wooden frame from Target is John's first response. The email where he called me a snowflake and told me he 'bangs his wife an animal [sic]'. I don't have any photos of John or Jane, but I do have this daily reminder that, at one time, he wanted to slap the shit out of me. While that's no longer the case and our email exchanges are far more convivial, I can't say that I've changed John's mind.

Maybe I am getting somewhere. Or maybe it's just hubris. In our last exchange, some months ago now, he'd sent a few abstracts of climate change denialism, work easily disproved with a quick Google search. I'd responded that he needed to check their provenance and think about the physics. His last reply provided something akin to ... hope, I think?

'I just had Don Pepe's for lunch. Will keep studying. Talk soon.'

✳ *Solomon Islands tribes sell carbon credits, not their trees*, p. **26**
Stanford prison experiment: Why the lead psychologist defended his infamous study to the end, p. **282**

STAYING FAITHFUL TO EARTH: A MEDITATION ON OUR PLANETARY KIN IN THE UNIVERSE

Ceridwen Dovey

It had been a long year, and I needed a holiday. I was so tired I didn't even know where to go. Just somewhere different, faraway, even exotic. I browsed online travel agencies and tourism sites until I found myself exactly where I wanted to be: at NASA's Exoplanet Travel Bureau.

'Scroll to take a trip outside our solar system,' they urged, so I did. Vintage travel posters in the style of romantic US national parks advertisements from the 1930s made me long to visit an exoplanet, a planet that orbits another sun. The posters are speculations created by designers who worked closely with scientists and futurists.

'Relax on Kepler-16b,' a poster of an exoplanet with two suns counselled, 'where your shadow always has company.' Using NASA's 360-degree visualisation tool, I visited a moon of Kepler-16b, where I could stand on the lunar surface, swivel and gaze at an artist's impression of the gas giant above me.

Then I popped over to TRAPPIST-1e, 'voted best "hab zone" vacation within 12 parsecs of Earth', and looked back at Earth's sun, which glows in the night sky of this exoplanet. I was there to see a rare alignment of the TRAPPIST-1 solar system's other six planets, just as travellers today, the bureau told me, travel far and wide to see eclipses.

I could easily imagine myself into the bodies of the humans

pictured on the posters, silhouetted against an exoplanet's landscape, but I also felt an urge to resist this identification. On the audio-enhanced guided tours, I became increasingly wary of this trope of tourism as a way of bringing the unknown, the unknowable, closer, as if with just a bit of imaginative effort you too could set foot on planetary kin far away, on planets that are deeply strange, sure, but still relatives of Earth and thus part of the same close-knit extended family of planets.

But are they really kin, these other planets? And does framing exoplanets as destinations of the deep future mean that we become a little bit less faithful to Earth, less inclined to worship at its altar of planetary exceptionalism?

It is a startlingly new discovery that there are more planets than stars in our galaxy. Even if early astronomers (like Kepler) intuited that other suns must have planets, we didn't have definitive proof until very recently that our solar system is not unique in consisting of planets orbiting a star. The first exoplanet was confirmed in 1992; the first exoplanet around a star similar to our sun was discovered in 1995. The latest count is over 5000 and growing. Discoveries have stacked up so fast that astronomers and astrophysicists who used to know each individual exoplanet by name now say it's impossible to keep track of those that exist in just one small part of the Milky Way, with thousands more expected to be found in the coming years.

The history of western astronomy is a history of the displacement of Earth as special. From Copernicus through Bruno to Galileo and beyond, each insight nudged us further away from being at the centre of the universe. No, the sun does not revolve around us – we are just one of several worlds that orbit it, and those other stars out there, those faraway ones, are actually other suns and have worlds around them, too. The perspective-altering consequences of what it means to live in a galaxy where planets are more plentiful than stars are still percolating through to us; there are so many exoplanets that a leading astrophysicist calls them 'commonplace', nothing but 'specks of dirt that collect around stars, like lint in a navel'.

Of course, the burning question that follows is: are any of those planets like Earth?

You may think that 'Earth-sized' or 'Earth-like' exoplanets – as often heralded in the media – are common, and habitable by humans if we could just figure out how to travel that far. They are not. The majority of detected exoplanets are so bizarre they make our planet look like the odd one out. To be fair, the current technologies of seeing and sensing (or simply deducing) the existence of exoplanets do bias discoveries in favour of exotic planets that are massive and very close to their stars. We cannot yet see or sense any planets truly similar to Earth because they would be so small, faint and slow-moving that it would be very difficult to block the light of their suns to know if they even existed – let alone to gather any clues about their surfaces.

When it comes to figuring out if an exoplanet is anything like Earth, and perhaps capable of sustaining life in the forms we know it, many scientists use the 'Goldilocks zone of habitability' metric. This considers whether an exoplanet is the right distance from its sun (not too warm, not too cold, like the porridge in the fairytale) to theoretically be able to have liquid water on its surface.

To a layperson like me, however, the Earth Similarity Index (though not universally endorsed by scientists) seems more useful because it includes so many different factors, in acknowledgement that it takes so much more than just the presence of liquid water to sustain life. Add to that the magnetosphere (which protects a planet from too much solar radiation), the right balance of water and rock, a moon just the right size, plate tectonics, a water cycle, the right kind of stable sun, a solar system that happens to be in a relatively quiet part of the Milky Way (and not in a dense galaxy that regularly emits lethal gamma-ray bursts), and an arrangement of sibling planets that is relatively settled.

To grasp just how many things had to go right for Earth

to be here – and us on it – is spooky. Certain essential factors immediately mean that entire swathes of the stars in our galaxy are discounted from hosting planets that could sustain life of any kind. And so, weirdly, in this abundance of exoplanets, I keep bumping up against Earth exceptionalism, almost holy in its reaffirmation of what a miracle it is that this planet exists in the way it does. If you don't believe me, the Planetary Habitability Laboratory has a poster you can put on your wall, titled 'Potentially Habitable Exoplanets', based on a ranking of exoplanets according to their Earth Similarity Index. Earth is ranked as 1.0. None of the others come close.

In a galaxy teeming with planets, the conditions under which our planet thrives – and the things it teems with, including us – seem to be unique.

I find this unsettling, especially as we begin to panic about what habitability means on Earth itself (habitable for who?). I don't want Earth to be the exception. I would much prefer it to be the rule. I want to feel a real and strong kinship with all the exoplanets, and I was pleased to learn that scientific opinion is slowly shifting towards the belief that it's almost arrogant to refer to them as exoplanets and not just as plain planets. After all, our solar system's planets are literally a drop in the universal planetary bucket, so why should all the thousands/millions/billions of others have to wear a prefix?

I could sense this same urge towards kinship and closeness emanating from the scientists who do the discovering of these planets, and it soothed me to observe that they could not resist knitting these affective, symbolic bonds across space and time to put Earth into a relationship with its planetary kin. I probably should not have been surprised by the fable-like, emotion-laden language of exoplanetary science. As the writer Vivian Gornick has noted, the inner lives of scientists and writers are remarkably similar, as both have to be open to using their imaginations to fathom what we don't yet know or may sometimes prefer not to know.

It feels significant that the exoplanet scientists who first described what they were finding (and what they were looking for) chose the language of family. 'It does something to you,' Didier Queloz, co-discoverer of the first exoplanet around a Sun-like star, for which he jointly won a Nobel Prize, has said. 'You defend it as if it were your own child.'

Poetic analogies in scientific language are nothing new, especially for things that are difficult or impossible for human eyes to see, being too tiny, too big, too far away or in the wrong part of the electromagnetic spectrum. So why does the language used by exoplanet scientists matter? It sets a tone, I think, especially in these early, establishing years. The language that follows will slip into these same thematics forever after.

To an outsider, this language is a useful clue to what's going on. When it sounds like scientists are telling fables, what types are they telling? What warnings or longings do those stories reveal?

Like me, you may start off being charmed by the warm, happy language of exoplanet kinship and close family ties. Exoplanets are said to belong to their parent – or mother – host star and are mostly named for the star they orbit. Some exoplanets have more than one star parent; they may orbit in multi-star systems, where the stars themselves have companions. Exoplanets usually have planetary siblings who orbit the same star.

Yet the deeper you go into the language – and the science – the darker it becomes, more and more a story of dysfunctional families, where some members see nothing of themselves in others and will do anything to be rid of them, even forcefully ejecting them from the family grouping. Ambivalence starts to burn through the analogies, and the language begins to describe not just the good, close stuff of familial bonds and loyalties (parent, companion, sibling, twin, cousin) but the worst kinds of things that can happen in a family (orphan, exile, rogue, evil twin, banishment).

We are learning now that planets are not as loyal to their parents as we might want them to be, and that most solar systems are rife with sibling rivalry. Planets fight for their parent's attention and compete for primacy. The Greek origin of the word planet translates to 'wanderer', named because of the independent movement of these dots of light in the night sky, always moving relative to the stars fixed behind them. The name is turning out to be prescient, for we must now get our heads around the fact that planets even within one solar system may migrate – from outwards in, and sometimes back out again – and like spinning tops they can jostle and collide and violently knock siblings out of the entire system. Any sibling unlucky enough to be ejected wanders the universe, starless. No parent star, no planetary siblings (though the weirdest and newest kind of these ejected planet-like objects, called JuMBOs, may travel in binary pairs).

Invisible family crimes may have enabled Earth's privilege in being exactly where it is, in its perfect place and position. Jupiter – a planet we like to think of as our protector, shielding us from asteroid or comet strikes in its current position – most likely ejected a sister or brother planet, or several, as it migrated inwards towards the sun.

A banished planet – let's think of it in the singular – is called a 'rogue'. Its existence has only become conceivable in the wake of all these revelations about other planetary families and the crashing down of accepted theories about how planets form around a star.

Our solar system – in its stable, neat family grouping with planets in almost perfectly circular orbits around a star – was generally assumed to be what any solar system out there would be like. Planets would gradually form according to the comforting organising principles of our system: small, rocky planets close to their sun; then gas giants further out, beyond the ice line; then the ice giants. And once formed, they would stay put, more or less.

Instead, what exoplanet scientists are finding is a cabinet of fascinating horrors, a mess of randomness, disorder and broken rules, and planets that are so vastly diverse in the forms they take as to be nonsensical. Yet the terms used to categorise exoplanets still refer to our solar system as the ur-system, almost desperately using analogy as reassurance. These alien exoplanets are described as terrestrial (indicating only that they're small and rocky), super-Earths or sub-Neptunes (planets that are much more massive than Earth but lighter than an ice giant like Neptune), and hot Neptunes or hot Jupiters (massive planets that orbit extremely close to their suns). This naming system implies that any differences between our planets and those planets are simply a matter of a few variances in size, mass and orbital period, and otherwise all else is equal. This could not be further from the truth. With current technologies, super-Earths and sub-Neptunes are the most common types of exoplanets being discovered out there in the galaxy, but they represent a size and type of planet that is entirely missing from our own solar system. No analogue for them exists in our system, in spite of their comforting names.

An exoplanet can be whatever it dares to be. It can orbit one star or two. It can orbit its star in 80 hours or 27 years. It can have many moons or none. It can be made partly of diamonds or lava, ice or steam, have the consistency of styrofoam or fluff. It can be a hot or cold 'eyeball' planet that is tidally locked with its sun (the way our moon is with us) so that half the planet never gets a break from staring at it, like an eyeball roasting in front of a fire. A few exoplanets – the zombies – orbit pulsars, which are dead stars.

Every new theory or finding unravels the next, and the next in turn. While preparing for my holiday to the exoplanet of my dreams, I read every recently published book for the general public on exoplanets, and each one made slightly different claims about what is definitely known or scientifically accepted, and what is still considered theoretical. What do the words familiar and exotic mean anymore, given that strange is the new normal in Exoplanet Land, and that Earth may just be the most exotic of all the exoplanets?

As soon as I learned of the existence of rogue planets, I knew exactly where I wanted to go on my once-in-a-lifetime adventure. But to which one? I kept reading, kept searching.

These rogues are also called 'orphans'. One word gives them a kind of bitter agency: they've gone rogue by choice. The other makes them victims of heartless fate: orphaned against their will, tossed out into the void.

Stuff you all, thinks the rogue planet. I don't need a parent sun, and my siblings hate me. I'm off. I never want to see any of you again.

I've lost my parent sun forever, thinks the orphan planet. I will never feel their light and warmth ever again.

The science of exoplanets is filled with speculation at every juncture, but it seems that orphan planets – which are believed to orbit through the galaxy, not around a sun – may outnumber the planets securely fixed in a family system around a star. There could be as many as 400 billion of them. Does it change your world view to know we may live in a universe of orphaned planets? Some scientists believe orphans could be likely candidates for hosting life, kept just warm enough by the internal radioactivity of their cores (as Earth is heated, in part, by radioactive elements at its core).

Exoplanets, it turns out, adhere closely to the Anna Karenina principle: all happy families are alike; each unhappy family is unhappy in its own way.

This ambivalence baked into the scientific language about exoplanets has helped me understand something that had previously puzzled me: why does the exoplanet scientific community mostly describe what they do as planet hunting? Not planet seeking or searching or discovering or finding, but hunting, as if exoplanets are prey they must get in the sights of their telescopes, like wild game caught in the sights of a rifle. Are they looking for planetary kin out there, or are they hunting down Earth's planetary rivals, picking them off, taking them out of contention for the ribbon for most habitable planet in the universe?

This term – planet hunting – was apparently first used in the 1980s at a fairly obscure NASA workshop on exoplanets when the field was still a bit embarrassing to mainstream astrophysicists and astronomers, when it lurked in the shadowy fringes of acceptable science, loosely aligned with the extraterrestrial fanatics who fantasised about communicating with intelligent life elsewhere. Maybe they chose to use the phrase planet hunting in part because of their own zeal and urgency: to become legitimate, they needed to find these things fast – they needed to hunt them down.

And they did, at first one by one like big game, each a trophy, a prize of enormous value. Then – thanks to the expanded abilities of space telescopes like Kepler and James Webb – the method became more like throwing an industrial-sized trawling net into the sea of stars and pulling up thousands of exoplanets.

For all the scientific analogies suggesting kinship and a close resemblance between our solar system and the others, I think that as the tally of exoplanets has risen, and their documented strangeness has shot off the charts, they have begun to scare us. Maybe that's why there is this slippage in the language, a tension between hunter and prey, kin and companion. Which is it? Do we want to hunt down and kill exoplanets, or love them as our own?

If exoplanets are becoming threatening to us in being so determinedly un-Earthlike, this is usually repressed in the scientific excitement over any new discovery. An exoplanet's otherness is quickly tamed, and so a small rocky planet with no hint of an atmosphere can be safely rebranded as somewhat similar to Earth simply for being small and rocky. The wild success of NASA's Exoplanet Travel Bureau posters and the public enthusiasm for new exoplanets when they're announced suggests that we feel only positively towards our planetary kin and immediately want to enfold them into the universal family.

But exoplanets are just so ... exo. They taunt us by being so out of reach we will never be able to see them for ourselves. They throw us back upon speculation, symbols, metaphors, visualisations, scientific analogies that work really hard but aren't quite right. For not even the scientists can see these exoplanets except through indirect means, learning to interpret complex graphs, charts, numbers, algorithms, and reams and reams of data. Exoplanets are so far away that they are frustratingly dry and boring visually – even the very few so-called 'direct images' show only a grainy dot. They are nothing like the gorgeous, lush, colour-soaked, precise (and highly processed, it must be said) images of galaxies and nebulae generated from the Hubble Telescope's data.

No wonder, then, that NASA is trying to reframe these grainy dots as glamorous, vividly imagined future tourism hotspots. In a world awash with spectacular imagery of the universe, the Exoplanet Travel Bureau ups the stakes by appealing to a person's fantasy of being there, setting one's own feet on alien ground. You cannot step on a nebula, no matter how beautiful it looks in the pictures. But a planet? As the popular astronomer Carl Sagan always insisted: planets are places.

In the end, I will admit, the pleasurable pull of the posters was too strong to resist. I gave in, let go of my suspicion of the whole project of commodifying exoplanets as holiday destinations. I found the perfect rogue planet to visit, PSO J318.5-22, a planet with no star, 'where the nightlife never ends'. I could see it all so clearly: how I would drink and dance in awe, get very cold, and become hopelessly homesick for Earth.

The benefit of spending time around far-flung members of a dysfunctional extended planetary family is that sometimes – for a moment, anyway – you remember to be grateful for what you have.

Until scientists can make clear images of the surfaces of exoplanets, which could take a very, very long time, their planetscapes will dwell most vividly in our imaginations. In our minds, perhaps exoplanets become not objects to hunt down but – like the whale in *Moby-Dick* – symbols of something we will never become familiar or comfortable with, no matter how hard we try: exoplanets as meditation, as philosophy, as constant reminder of Emily Dickinson's poetic and haunting line, 'This World is not Conclusion'.

An astrophysicist friend told me recently that stars are simple; a star with the same mass, bulk, composition and age as the Sun is almost identical to it in every other respect. Yet planets, he said, are never simple. They are endlessly diverse, because they are shaped by their own histories, environments and interactions – by their relationships with their parents and siblings. His motivation in searching for exoplanets has nothing to do with grandiosity, saving our civilisation or finding a backup Earth, as people often wrongly assume. It's about contemplating places so strange and far away they cannot ever be fully seen or known, 'a practice of going out and staying out, a way of holding the scale of the world in my mind'. On the good days, that kind of close attention to exoplanets allows for a fuller and deeper faith in Earth, for being the planet that Goldilocks herself might describe as just right.

✱ *A freediver finds belonging without breath*, p. **69**
Sounds of the slow-rolling sea, p. **175**
Deep time encounters in the garden, p. **287**

WHALE TALK

Drew Rooke

When Shane Gero first travelled to Dominica in January 2005 to study the sperm whales who live in the sparkling, squid-rich waters that surround the Caribbean island-nation, he did not anticipate how transformative the trip would be. 'I thought we were just going to go there for that one season,' he says, laughing.

Back then Gero was a young masters student at Dalhousie University in Canada. He had loved whales and had dreamt of becoming a marine biologist ever since he was a small child and now he was living out his dream.

Accompanying Gero were his supervisor, biologist Dr Hal Whitehead, and several other graduate students – all living and working together aboard *Balaena*, Whitehead's 40-foot cutter-rigged sailboat. They had chosen the waters around Dominica as a research site after learning that there were many more young whales in family units there than elsewhere. 'And I wanted to study what it was like to grow up as a sperm whale,' Gero says.

As it turned out, there was no better location in the world to do so. For 41 consecutive days, the group sailed alongside one family of sperm whales – nearly triple the longest amount of consecutive time Whitehead, who was one of the foremost sperm whale researchers in the world, had spent in the company of these ocean nomads. 'I'd love to tell you that it's my acumen as a biologist, but we just kind of found this amazing place to study these amazing animals,' Gero says.

To continue his research, Gero founded the Dominica Sperm Whale Project and began making regular trips to Dominica to study its resident cetaceans. He hasn't stopped: these days, he spends roughly three to four months every year in the field to continue working on his long-running research program, which over the last two decades has collected the largest sperm whale data repository in the world.

'We've seen calves born and have followed them through weaning. Each time we go back we notice particular individuals have grown up, or that they suddenly have new marks on their bodies. They're really part of our lives, kind of in the same way as a distant colleague or a friend that you only see once or twice a year,' says Gero.

Now he hopes to further deepen his connection with these gentle giants of the deep – by cracking the code of their complex communication system and maybe, just maybe, talking with them.

Whales find their voice

Cetaceans appeared on Earth roughly 60 million years ago when their land-dwelling ancestors began to venture further from the shore, evolving traits to survive their new, aquatic home.

Over time their front limbs were replaced with flippers, while their obsolete hind limbs disappeared. Their tails grew wide and fluked, their bodies became insulated with a thick layer of blubber, and the haemoglobin levels in their blood swelled to enable more efficient storage of the precious oxygen they inhaled at the surface. In addition, this new order of mammals developed the ability to communicate with each other underwater.

Vocalisations are not uniform among different species. Toothed whales, such as sperm whales and dolphins, produce short sequences of clicks and whistles known as 'codas', by releasing a high-pressure blast of air which passes through a vibrating structure in their nose known as 'phonic lips'.

In the case of sperm whales, these sounds are one the loudest single-source noises on Earth, reaching 200 dB – equal to the 1967 launch of the Saturn V space rocket, and loud enough to burst a human eardrum – and help with echolocating and hunting prey in the deep, dark depths.

In contrast, the melodic songs of baleen whales such as humpbacks are produced as air passes from their lungs and through a U-shaped fold of tissue, causing it to vibrate. The resulting sound then resonates in an inflatable organ called the laryngeal sac.

Speaking to scientists?

Communicating with these highly vocal and social marine mammals has been a long-held pipe dream of scientists – one that has historically attracted some eccentric personalities.

Most notable is American neuroscientist John Lilly, who in the late 1950s established the Communications Research Institute (CRI) on the shores of Nazareth Bay, on the eastern side of the island of St Thomas in the Caribbean Sea.

Up until that point, Lilly had conducted most of his pioneering – albeit highly unethical – research into brain function, behaviour and consciousness using monkeys and cats as experimental subjects. He had been interested in studying cetaceans since 1949, when he and two colleagues extracted the brain of a pilot whale which had fatally beached in Maine; its large size, he believed, indicated a high degree of cognitive complexity, equal to – or perhaps greater than – our own.

Lilly's interest deepened throughout the 1950s when he made several visits to Marine Studios, an oceanarium in Florida, to conduct experiments on some of the bottlenose dolphins held captive there and heard for the first time their complex vocalisations.

These experiments proved fatal for many dolphins. But they proved life-changing for Lilly, inspiring him to build 'the world's first laboratory devoted to the study of the intellectual capacities of the small, toothed whales', as he wrote in 1959 to a friend at NASA, which soon thereafter provided funding for the CRI in the hope that its research would help lay the foundations for communicating with extra terrestrial intelligence.

As well as studying the creaks, clicks and whistles of the captive dolphins at the CRI – which, in Lilly's mind constituted a language he termed 'dolphinese' – the research program at St Thomas was aimed at teaching these animals a human language: 'a primitive version of English', as Lilly wrote in his 1961 bestselling book, *Man and Dolphin*.

This would require, Lilly reasoned, 'constant and continuous attention and awareness of detail'. Therefore, one of his assistants, Margaret Howe, lived full time with a dolphin named Peter on the flooded and waterproofed second floor of the building, attempting to teach him basic words and phrases. Over time, Peter managed to produce sounds that bore some resemblance to English words (much like some parrots can) in exchange for fish treats, but made little other progress.

These meagre results contributed to the CRI – and, by extension, Lilly – losing its scientific credibility and much of its funding. This decline rapidly accelerated when reports emerged that Lilly had dosed his captive dolphins with LSD to hopefully help them learn human language skills – and that Howe had manually relieved Peter of his sexual desires, which were becoming increasingly disruptive to the institute's interspecies communication work.

By the end of the 1960s, the CRI had closed down. But Lilly continued his quest to communicate with cetaceans all the way up until his death in 2001, believing that the benefits of succeeding would be extraordinary.

As he wrote in 1978, 'Let us learn to communicate with the ancient macrobiocomputers of the Cetacea and learn something of the complexities of their computational capacities. Such communication may enrich our lives beyond anything that we have heretofore conceived and may open up possibilities for the future evolution of man beyond his present limits.'

Whale phonetics

Founded in 2020, the Cetacean Translation Initiative – Project CETI, for short – agrees with Lilly about the huge worth in learning to communicate with cetaceans. But it has taken a very different approach in its quest to achieve this.

Instead of using hallucinogenic drugs on captive whales and trying to teach them English, it is using the combined skills and experience of biologists, linguists, cryptographers, acoustic engineers, roboticists, computer scientists and artificial intelligence experts to try to understand their ancient communication systems.

Gero is the lead biologist on the project, and earlier this year he co-authored a study which he believes marks a significant step towards breaking the interspecies communication barrier.

Published in *Nature Communications*, the study sought to shed new light on the communication system of sperm whales and was based on more than 8700 sperm whale codas.

This enormous dataset had been collected as part of the Dominica Sperm Whale Project between 2005 and 2018 using towed listening devices plus sound tags placed on individual animals.

Analysing it manually – by trawling through reams of printed spectrograms, as scientists once had to do – was practically impossible. Instead, Pratyusha Sharma, a PhD student in the Computer Science and Artificial Intelligence lab at MIT and the lead author of the study, used a combination of statistical and machine learning methods to look for patterns and features within the whales' click sequences.

This proved groundbreaking. It revealed, firstly, that sperm whales make fine-grained adjustments to their codas depending on the conversational context, such as adding an extra click – 'kind of like a suffix', explains Sharma – or varying the duration of their calls.

Secondly, the analysis found that the whales freely combine these variations to construct a repertoire of distinct vocalisations far larger than was previously believed. 'And the interesting thing about combinatorial communication systems like this one is that there are not that many examples of it in the world,' Sharma says. One of the only other examples is human language. 'We have alphabets that combine to form words and words that combine to form sentences, and that's how we can use finite sounds to … express infinite meanings.'

The researchers catalogued these newly discovered variations into what they called a 'sperm whale phonetic alphabet' – like the International Phonetic Alphabet for human languages – which they believe provides a foundation for future research into the semantics of whale calls.

Acoustic exchanges

Key to the next stage of research are what are known as 'interactive playback experiments', which are already being conducted with some other whale species.

Most notably, in November 2023 a team of researchers from the SETI Institute, University of California Davis and the Alaska Whale Foundation, published the results of an interactive playback experiment they conducted with an adult female humpback whale named 'Twain' in south-east Alaska two years' earlier.

The experiment involved broadcasting via an underwater speaker a high-quality contact call, known as a 'whup' call, which had been recorded the previous day from a group of nine humpbacks. This attracted the attention of Twain, who approached and circled the team's boat and began responding with her own call. This 'acoustic exchange', as the researchers described it, continued for 20 minutes.

Later analysis of the recording revealed that Twain was, as the study said, 'actively engaged in a type of vocal coordination [with our playback system] ... she was also exhibiting changes to both arousal and valence during the encounter'.

According to the study's lead author, Dr Brenda McCowan, this marked an unprecedented step in interspecies communication research. 'We believe this is the first such communicative exchange between humans and humpback whales in the humpback "language".'

Uncertain understanding

Not everyone agrees with the findings to date. There's some contention among scientists about the significance of these recent studies with sperm and humpback whales, as well as about the possibility of interspecies communication more broadly.

In fact, when I ask Rebecca Dunlop – an associate professor in physiology at the University of Queensland who has been researching humpback bioacoustics for over two decades – if she thinks that we are on the cusp of being able to communicate with cetaceans, she chuckles and says bluntly, 'Nope.'

Dunlop acknowledges that interactive playback experiments can help determine the function of whale calls. But she dismisses the idea that scientists are in some way conversing with a whale if they broadcast a call and receive an engaged response, as happened with Twain. To claim this implies that 'the whale heard them and then decided to say something back, like there was some cognitive decision making going on, which I think is a bit of a stretch'.

Instead, she sees the interactive playback experiment with Twain as demonstrating 'an animal responding to a sound, as it's pre-programmed to do'.

Dunlop is also sceptical of some of the claims made by the Project CETI team in their paper about a sperm whale phonetic alphabet. While she accepts that sperm whales' clicks are very complex, she doesn't believe that they can be equated to human language.

'Humans have evolved language with syntax, and we can change the meaning of a sentence by just putting words in a different order. How we use language is highly, highly complex, and other animals – as far as we know – are nowhere near that level complexity. They can use sounds to mean certain things. But to say that that's a language is I think one step too far.'

Dr Jenny Allen – a biologist from the Bio-Telemetry and Behavioural Ecology Lab in the Department of Ocean Sciences at the University of California, with more than 15 years experience researching humpback whales – has other criticisms.

There is, Allen says, 'tremendous value' in conducting interactive playback experiments, adding that scientists have been conducting them with birds 'for ages'. But to frame the one conducted with Twain as a kind of primitive conversation between humans and a whale makes the mistake of 'pushing these animals through a human-shaped hole'. 'Animals communicate to each other in such a variety of ways that are so different to what we know.'

Allen is also critical of how the Project CETI team characterised sperm whale vocalisations, noting that while the previously undiscovered features are 'really fascinating', the idea of a phonetic alphabet is 'beyond the scope of the study'.

'It means that each individual component doesn't have a meaning. And we can't really say sperm whale codas by themselves have no meaning.'

Speaking more generally about the quest to translate whale vocalisations and talk with them, she says, 'I worry that we're so busy looking for signs that point to other animals being like us, rather than looking at the similarities that we do find and asking, "What does it mean for that species? What does it say about evolution?"'

But Josephine Hubbard, postdoctoral researcher in the Animal Behaviour Graduate Group at the University of California Davis, who co-authored the study involving Twain, says that her team was 'careful not to anthropomorphise our interpretation of the data, which is why we were very strict in how we define a communicative exchange'. She also emphasises that there is 'robust evidence' the female humpback was engaged during the experiment. 'And we can debate about whether it was a conversation or not, but I think what's worth highlighting is the fact that we are using and trying to promote these interactive playbacks.'

Likewise, Pratyusha Sharma of the Project CETI group disagrees with the criticisms levelled at the study she led. 'We don't make any claim that this formal communication system of sperm whales is like human language. But there are aspects of it that are similar.'

She also points out that the word 'alphabet' is widely used by scientists to characterise many complex structures, including DNA, and cites the example of the hieroglyphs to point out that in some alphabets, even the smallest units do indeed carry meaning.

Adding to this response, Gero insists that the Project CETI team isn't 'trying to create a hierarchy which ends in humans. But I do think that we're at a stage now where we can ask more detailed questions about animal communication.'

More broadly, he believes all of us should keep an open mind about the intellectual capacity of nonhuman species. 'I think we do a disservice to whales if we assume they have some kind of stimulus-output quality, when we know that they have a brain that rivals ours at least in terms of the capacity for cognition.'

The desire to communicate

Many books have been written over the years that envision what it would be like if humans could communicate with other animals.

Laura Jean Mckay's science fiction masterwork, *The Animals in that Country*, in which a new virus emerges whose chief symptom is that its victims are able to understand what animals are saying, is one of the wildest examples to date.

If science fiction does indeed one day become science and we manage to understand whales and how to communicate with them, I wonder what we might say?

I ask Sharma this question. After a few moments of trying to find an answer, she says, 'You know, I would not say anything. I would just want to listen and hear everything they have to say about their world.'

Gero, likewise, sees Project CETI as being a 'listening project' – which will hopefully enable us to better understand what's important to whales and thus help conserve them.

'I feel very strongly as a scientist and just someone who spent a huge amount of time with these whales that we need to do excellent science and ask what's going on in their world. But then, importantly, we also need to deliver on that and ask ourselves, "Well, what are we going to do about it?"'

✱ *Insect consciousness*, p. **87**
Why can't we remember our lives as babies or toddlers?, p. **204**

GAZA: WHY IS IT SO HARD TO ESTABLISH THE DEATH TOLL?

Smriti Mallapaty

Since war broke out in the Gaza Strip almost a year ago, the official number of Palestinians killed exceeds 41 000.* But this number has stoked controversy. Some researchers think it is an underestimate, owing to the difficulties of trying to count dead people during conflicts. Other sources say it overestimates the number of casualties. The count comes from the Palestinian Ministry of Health – Gaza, the main institution counting mortality in the region.

It's important to track fatalities during wars – and to estimate overall mortality – to hold warring parties accountable and to advocate for the protection of civilians, says Zeina Jamaluddine, an epidemiologist at the London School of Hygiene & Tropical Medicine. The number of deaths also informs discussions around when to officially declare that a situation involves famine.

In the heat of conflict, the first way to count fatalities is to tally up the number of dead people. But capturing the number of deaths in the densely populated urban centres of Gaza presents unique challenges, says Emily Tripp, director of Airwars, a non-profit watchdog based in London that counts casualties in times

---

* This article was published on 24 September 2024. Since then, the death toll has risen considerably. Factors such as the targeting of hospitals by the Israeli Defence Force have further limited the ability for accurate estimates. The UN Special Committee has stated 'Israel's warfare in Gaza is consistent with the characteristics of genocide'.

of conflict. 'What we've seen in Gaza is entire families just being completely wiped out,' says Tripp. That means it can be hard to recover bodies, or there is no one to report them dead, and so deceased people will be missed in counts.

Only when the conflict ends or eases can researchers begin the work of getting more robust estimates of overall mortality through surveys, modelling and statistical tools, they say.

Counting deaths

The Palestinian Ministry of Health – Gaza updates the death count almost daily, and has published five detailed lists of people who have died, including, where possible, full names, national identity numbers, age and sex.

In the first few weeks of the war, the ministry recorded deaths that were reported at hospital morgues, using a system well established in Gaza. The early lists were detailed enough that researchers, including Jamaluddine and her colleagues, found that the mortality data and daily reporting followed similar trends to the deaths among staff members working at the international agency providing humanitarian assistance in Gaza, the United Nations Relief and Works Agency for Palestine Refugees in the Near East (UNRWA). Airwars also found that the ministry's data were in accord with its own.

But when medical centres came under attack, the ministry's ability to count fatalities was affected, says Jamaluddine, who noticed the change after the attacks on Al-Shifa Hospital in Gaza City in November. 'It became harder and harder as the war broke out.' The World Health Organization reported that as of 17 September, fewer than half of Gaza's 36 hospitals were still operational, and that those that were were providing only limited services.

Since then, the ministry has had to increasingly rely on deaths reported outside hospital morgues by the Palestinian Civil

Defence, the Palestine Red Crescent Society or ministry staff, says Omar Hussein, director of the emergency operations centre at the Ministry of Health in Ramallah. In areas not accessible to these organisations, particularly in northern Gaza, the ministry verifies deaths reported by close relatives, he says.

The ministry's most recent list of people who have died, published in July 2024, includes complete information on approximately 28 000 people. The ministry reports that there are still some 7000 people for whom they don't have complete information, owing to the difficulties in identifying them. Many deaths are still unaccounted for because the people are buried under rubble, says Hussein.

Some deaths could also have been missed because friends and relatives don't want to report them, or don't know a person has died. 'Nearly everyone in Gaza right now is displaced, so they don't know where everybody else is, and although most people have cell phones, cell coverage is intermittent,' says Patrick Ball, a statistician and director of research at the non-profit Human Rights Data Analysis Group in San Francisco, California. The medical system could also be too overwhelmed to report them.

But Mark Zlochin, an unaffiliated researcher in Pardes Hanna-Karkur, Israel, who has analysed the ministry's fatalities data, says there is no way to verify the number of deaths for which the ministry has incomplete information. Although he thinks the specific deaths reported through the ministry hospital morgue system – a large portion of the 28 000 deaths recorded in the July list – are probably true casualties, he thinks the ministry's total reported deaths is probably an overcount. Deaths reported by relatives could include people who have just gone missing, and young children and older individuals who died for reasons not related to the war, he says.

Overall, the ministry's records are 'pretty good' because they include detailed information for most fatalities, and that transparency provides credibility, says Michael Spagat, an economics researcher at Royal Holloway, University of London, in Egham, and chair of the non-profit advocacy group Every Casualty Counts in London.

Besides the ministry, other groups tracking casualties in the region include the UNRWA, which collects information on its staff members who have died; independent organisations such as Airwars and Uppsala University's Uppsala Conflict Data Program (UCDP) in Sweden, which use ministry data and collect information from local news agencies, non-governmental organisations and local groups that keep track of deaths and report them on social media.

The UCDP recorded more than 36 000 fatalities in Gaza between October and July, and, in upcoming revisions, that sum will probably be closer to the 39 000 reported by the ministry during the same period, says Nanar Hawach, a researcher at the UCDP.

Estimating deaths

Given the uncertainty of counting fatalities during conflict, researchers use other ways to estimate mortality.

One common method uses household surveys, says Debarati Guha-Sapir, an epidemiologist who specialises in civil conflicts at the University of Louvain in Louvain-la-Neuve, Belgium, and is based in Brussels. A sample of the population is asked how many people in their family have died over a specific period of time. This approach has been used to count deaths in conflicts elsewhere, including in Iraq and the Central African Republic.

The situation in Gaza right now is not conducive to a survey, given the level of movement and displacement, say researchers. And it would be irresponsible to send data collectors into an active conflict and put their lives at risk, says Ball.

There are also ethical concerns around intruding on people who lack basic access to food and medication to ask about deaths in their families, says Jamaluddine. Surveys will have to wait for the conflict to end and movement to ease, say researchers.

Another approach is to compare multiple independent lists of fatalities and calculate mortality from the overlap between them. The Human Rights Data Analysis Group used this approach to estimate the number of people killed in Syria between 2011 and 2014. Jamaluddine hopes to use the ministry fatality data in conjunction with those posted on social media by several informal groups to estimate mortality in this way. But Guha-Sapir says this method relies on the population being stable and not moving around, which is often not the case in conflict-affected communities, including Gaza.

In addition to deaths immediately caused by the violence, some civilians die of the spread of infectious diseases, starvation or lack of access to health care. In February, Jamaluddine and her colleagues used modelling to make projections of excess deaths due to the war and found that, in a continued scenario of six months of escalated conflict, 68 650 people could die from traumatic injuries, 2680 from noncommunicable diseases such as cancer and 2720 from infectious diseases – along with thousands more if an epidemic were to break out. On 30 July, the ministry declared a polio epidemic in Gaza after detecting the virus in sewage samples, and in mid-August it confirmed the first case of polio in 25 years, in a 10-month-old baby.

Survivor stories

The longer the conflict continues, the harder it will be to get reliable estimates, because 'reports by survivors get worse as time goes by', says Jon Pedersen, a demographer at !Mikro in Oslo, who advises international agencies on mortality estimates.

Even after the violence winds down, many factors will influence the reliability of mortality estimates. For example, if the war ends with, essentially, a military occupation by Israel, people who have been hiding deaths from officials might continue to do so out of fear, says Ball. Local trust in the institutional structures set up after the war matters, he says.

But several factors mean that it should be possible to reconstruct the death toll, says Leslie Roberts, an epidemiologist who specialises in measuring mortality and human-rights abuses, and holds an emeritus position at Columbia University in New York City. For example, Gaza's population is highly educated and the region has strong social networks, making it easier to track where everyone is.

✱ *Ethical problems continue to plague biometric studies of Chinese minority groups*, p. **215**
From hypnotised to heretic: Immunising society against misinformation, p. **228**

SOUNDS OF THE SLOW-ROLLING SEA

Sara Webb

On June 2023 the internet lit up with excited physicists hinting they had found something groundbreaking. Imaginations ran wild: had we heard from aliens? Broken general relativity? Uncovered a hidden dimension in the universe? When the secret spilled, the truth was as good as the speculation: physicists had found evidence of a gravitational wave background.

This was an impressive feat. These ripples in the fabric of space-time are hard to spot; even booming gravitational echoes from black holes colliding fade to whispers by the time they reach Earth. Hearing the 'hum' of a low-frequency gravitational wave background required a galactic-scale detector – made up of dead stars.

So how did they do it? And more importantly, why are physicists saying that the most exciting part is yet to come? Come with me through space and time to find out if this discovery could change the way we study and understand the cosmos.

Gravitational waves 101

Wait – what are gravitational waves, anyway? Consider this your primer before we get to the world-changing stuff.

It all began, of course, with Albert Einstein.

In 1915, Einstein published his general theory of relativity, describing that the force of gravity we measure is due to the bending of space and time. Einstein was the first to propose that space and time were intertwined, and that they could be described as acting like a fabric. Picture a trampoline. If you put a bowling ball in the middle, the fabric is stretched and pulled downwards as the bowling ball's mass creates a dip. Now replace the bowling ball with a marble. It will create a much smaller dip on the trampoline. Those dips represent gravitational wells in space and time. Everything with mass – including you and me – creates these gravitational wells.

Einstein moved quickly. By 1916 he had used this new knowledge of gravity to postulate the existence of gravitational waves: ripples through space and time, caused by the movement of mass. Put your bowling ball back on the trampoline, but now move it up and down (that is, accelerate its speed), and you'll start to see ripples as the trampoline fabric is slightly bent and stretched. The physics that causes these ripples, as Einstein explained, involves the loss of energy through gravitational radiation.

It was an incredible theory and one that no one had really considered before: the fact energy could be lost through gravity. But Einstein came to suspect it could never be proven. You see, although anything with mass accelerating technically creates gravitational waves, they are so vanishingly small that detecting them is difficult. Yet in the decades after Einstein, physicists realised that detection was not impossible: we would just need stunningly precise instruments, and very energetic gravitational waves caused by enormous masses – for example, the collision of two black holes.

Here's the thing: Einstein famously didn't believe black holes could exist.

If physicists of the last century had just left it there, we wouldn't have even looked for gravitational waves, and you wouldn't be reading this article ... but we did, and you are.

Gravitational waves 201

Almost exactly a century later, in 2015, we found them – using the 4-kilometre-long lasers of the Laser Interferometer Gravitational-

Wave Observatory (LIGO), based in the US. LIGO uses instruments called Michelson interferometers, which split a beam of light, send the two beams travelling out the same distance in perpendicular directions, then reflect the beams back to join again. If nothing in the system changes, the same light patterns occur. However, if one arm of the detector is slightly stretched or slightly contracted – say, as a gravitational wave travels through the Earth – the light pattern changes.

LIGO's first detection was of gravitational waves formed when two black holes merged 1.3 billion light years away. As the waves rippled through space-time from this energetic event, Earth was stretched the slightest amount – less than 1000 times smaller than a proton – but remarkably, LIGO's interferometers could detect it.

In under ten years, we've detected more than 85 other mergers of massive objects. But these are all discrete events: single massive booms echoing across the universe, mostly from black holes colliding. These events create something called high-frequency gravitational waves, which are short and sweet and easy(ish) to detect. However, astronomers have theorised for decades that low-frequency gravitational waves should exist, hiding in the background of the universe.

Remember that any mass accelerating creates gravitational waves. With this logic we'd expect the universe to be awash with these ripples, a cacophony of movement adding up to a background hum. If we could hear it, these background ripples might even hold clues to the universe during the first moments of its existence.

(Pulsar) timing is everything

To detect a background of gravitational waves in a completely different range of frequencies, LIGO wasn't going to cut it. We needed a different approach: an epic, astronomical-scale experiment.

Instead of using lasers, scientists turned to cosmic lighthouses. When massive stars die, they generally form either a black hole or a neutron star.

Neutron stars are incredibly dense, rapidly rotating and highly magnetised objects. Most are only 20 kilometres across, but weigh at least the mass of our Sun. These little objects are hard to find; we can't generally see them in optical light, but we can spot them in radio light, because as they spin they emit a beam of electromagnetic radiation from their poles, mostly in radio wavelengths.

The first neutron star was discovered in 1967 by Jocelyn Bell Burnell, a PhD candidate at Cambridge University in the UK, who was analysing data from the newly built Cambridge radio telescope when she noticed peculiar pulses, from the same patch of sky, day after day. Bell Burnell had discovered the first pulsar.

After 57 years, we've discovered over 3000 of these cosmic lighthouses, and we're using them to map space and time in our galaxy. Pulsars are incredibly predictable and so astronomers use them as precision timing instruments. The pulses we see from Earth won't change unless something has interfered with them – just like LIGO's laser beams. What could possibly interfere with a pulsar's signal? Perhaps very long gravitational waves stretching space between us and them?

This brings us to the International Pulsar Timing Array (IPTA), which consists of four research teams in Europe, North America, India and Australia. Each is searching for low-frequency gravitational waves by monitoring the arrival times of pulses of more than 100 millisecond pulsars, thousands of light-years apart. While ground-based detectors can detect gravitational waves from 10 hertz to 10 kilohertz, the sheer size of the IPTA is sensitive to gravitational waves between 10^{-10} Hz to 10^{-6} Hz.

In 2023, it was the results of the IPTA that sent those shockwaves around the world, when multiple teams reported the first evidence of a gravitational wave background.

Australia's contribution is called the Parkes Pulsar Timing Array (PPTA) project.

'Murriyang [the 64-metre-wide CSIRO Parkes radio telescope in New South Wales] has been observing millisecond pulsars to detect gravitational waves since 2004, and as a result is the world's longest-running pulsar timing array experiment,' says Daniel Reardon, an astrophysicist at Swinburne University, Melbourne, and part of PPTA.

Reardon led one of those papers that broke the internet in June 2023, in which the PPTA team found that their analysis of 18 years of data was consistent with an isotropic gravitational wave background. What this means is that in all directions, a similar signal of a gravitational wave background is present.

But the data is still a bit of a mystery. 'We don't know the source of the gravitational waves yet,' Reardon points out.

One theory is that the gravitational wave background is created by supermassive black holes, which are found at the centres of most galaxies, including our own. Each is millions to billions of times the mass of our own Sun, and one of the greatest mysteries in physics currently is exactly how and when these giant monsters formed. The smaller, merging black holes that we see with LIGO are remnants of massive stars exploding at the end of their lives. Supermassive black holes, however, can't be explained by stellar death alone. Even more curious is that we often see galaxies merge in our universe – and when they do, we'd expect their supermassive black holes to merge too. In theory, these events would help create the gravitational wave background.

But according to Reardon, the new detections aren't exactly as expected. 'If it is the symphony of all binary supermassive black holes in the universe, then it's a little louder than expected and has hints of other interesting properties,' he says.

Potentially these gravitational waves occurred earlier in the universe than we first expected. This would be groundbreaking, as we don't know how and when these supermassive black holes formed in the first place. Understanding when they started to merge could help us unlock the secrets of their formation.

Beyond supermassive black holes, astronomers also theorise that a gravitational wave background could have origins in the Big Bang itself.

In principle there might be gravitational waves left over from cosmic inflation, less than one second after the very beginning of time, when the universe suddenly expanded faster-than-light. Many physicists believe that this expansion magnified quantum fluctuations – tiny, random energy changes – which then became the seeds of all the large-scale structure of the universe today. 'It's possible that gravitational waves created from quantum fluctuations could have been amplified by inflation into low-frequency gravitational waves,' Reardon says. 'But for it to be observable we'd need certain conditions to be satisfied.'

These conditions relate to the fundamental physics of the pulsars themselves, and those conditions aren't currently satisfied. It's much more likely the signal we recently detected is from merging supermassive black holes, just occurring slighter earlier in the universe then we expected.

Old universe, new frontiers

Gravitational waves – and particularly a gravitational wave background – are a game-changer for astronomers, who have the challenging job of determining the very nature of the universe: its shape, size and make-up. For a long time this task was limited by our ability to detect light, which is historically the dominant way that we receive astronomical information. Over the centuries, improving technology has allowed us to capture more and more light from the cosmos, beyond optical light and across the range of the electromagnetic spectrum. This light has given us insight into different time periods and different kinds of events: infrared light, for example allows us to peer back billions of years ago and see the very first galaxies, while ultraviolet light allows us to trace star formation within galaxies.

One of astronomy's most revelatory discoveries was made in the early era of radio telescopes, only a few years before Bell Burnell's work on pulsars. Across the Atlantic in New Jersey, United States, two scientists spotted the earliest light we can see: the cosmic microwave background (CMB).

It was 1965, and Robert Wilson and Arno Penzias were employed for Bell Laboratories working with the Holmdel Horn Antenna. The Horn was sensitive to microwave light, a subsection of high-energy radio waves, and Wilson and Penzias were exploring the instrument's potential uses for radio astronomy.

But their measurements were humming with excess noise, no matter which direction the Horn was pointing. They worked tirelessly to eliminate any possible sources of interference, even rebuilding parts of the Horn and cleaning away pigeon poop, but to no avail. In the end, Wilson and Penzias concluded that the noise was likely coming from beyond our galaxy.

Only 60 kilometres away at Princeton University, physicists had theorised that microwave light could be found left over from the beginning of the universe. We know now that the noise detected by the Horn was exactly that: leftover light from approximately 380 000 years after the Big Bang, now scattered evenly across the entire observable universe and stretched out to microwave wavelengths.

In the years since, studying the CMB has been crucial to the study of the early days of cosmic existence, providing the best data we have about the infant universe before stars and galaxies formed.

While the cosmic microwave background is based on very different physics to the gravitational wave background (GWB), it's possible that the GWB also has the similar potential to revolutionise astrophysics. Monash University PhD candidate Rowina Nathan, who was involved in the PPTA gravitational wave study, says the

GWB could yield information that is inaccessible with traditional electromagnetic astronomy.

'Due to the speed of light, when we look at things far away, we are seeing them as they were a long time ago,' she explains. 'Presently, the furthest we can see is the cosmic microwave background, as that is when the universe became transparent to light. The gravitational wave background may be a way to "see" past this, to a time before the universe was transparent to light.'

Nathan adds that the new GWB results are also tantalising astronomers with clues of new and exciting physics.

'A key idea in cosmology is that the universe is homogenous and isotropic – basically that when you zoom out far enough the universe is approximately the same in all directions,' Nathan says. 'If we see significant variance in the background, say it's stronger in one direction than another, this could disprove this, having significant implications for our understanding of the formation of the present-day universe.'

If that statement doesn't make you excited, it should! If the universe isn't the same make-up or structure in all directions, we might need to take a lot of physics back to the drawing board. Say we don't see merging supermassive black holes in a particular direction – could that mean galaxies are made up differently in different places? This astronomer thinks it's extremely unlikely, but a very exciting possibility nonetheless.

The gravitational wave background is a relatively new kid on the block, but one to keep our eyes on – because the IPTA teams are nowhere near finished with their work.

'In the next few years, we can look forward to the IPTA combining all 115 pulsars to calculate a higher statistical significance of the GWB detection,' Reardon tells me. This means the teams will have more data to help them confirm the signal, as well as adding in continuing observations.

I, for one, can't wait to see if the universe will surprise us.

✱ *Mathematical transformations: The power of vectors and tensors*, p. **20**
Staying faithful to Earth: A meditation on our planetary kin in the universe, p. **148**

JOCELYN BELL BURNELL, MAE JEMISON, CAROLYN BEATRICE PARKER

Alicia Sometimes

Jocelyn Bell Burnell

If we assume we've arrived:
we stop searching, we stop developing

proud pulsars
residue
of massive stars
      collapsing into
      a neutron star
strobing supernova
      spherical and dense
      size of a bustling city
more mass than our Sun

beams of electromagnetic radiation
whirling from its magnetic poles
      recurring signals
      palpitating refrains from space
—first observed in 1967
by Jocelyn Bell Burnell

and Antony Hewish—
Someone takes a photo:
Look happy dear, you've just made a discovery!

millisecond pulsars can siphon matter
and momentum from its companion

in 1974 Hewish receives
the Nobel Prize in Physics
but not Bell Burnell
she wins other weighty awards
becomes a Dame, donates
substantial sums to further
those ignored in science

pulsars radiating light
in a manifold of wavelengths

Mae Jemison

Hailing frequencies open

Physician/Astronaut/Chemical Engineer
beginning of many careers
        ready to launch vertical
into history
Numbers count:
        in 1992 Jemison logged
190 hours 30 minutes 23 seconds
in space on the STS-47 mission
Space Shuttle *Endeavour*
        126 circles in stable transit
dizzying in thin wings above
blue and white swirls of Earth

Mission Specialist 4 knows
she belongs here *as much as any*
speck of stardust, any comet, any planet
researching weightlessness, bone cells
threading connections and insights

Images matter:
carries with her a picture—Bessie
Coleman, aviator Queen Bess—
brings a poster of Judith Jamison
dancing along for the ride
        (performing Alvin Ailey's 'Cry'
—the half-moon of her arm
declaring a path beyond the void)

as a child, she watched Nichelle Nichols
Nyota Uhura levelling orders
        on the USS *Enterprise*
Mae Jemison in low Earth orbit
creating a dynamic first of her own
the gravity of this moment
the thrill of this adventure
        opening up space
        where it wasn't before

Carolyn Beatrice Parker

It is a silvery metal
        in a dark room
blue-skewed glow
excited by decay
Does Parker
        hold the polonium in her hands
does she ever breathe it in?

Working government top-secret
with this radioactive element
—The Dayton Project
part of the Manhattan—
research and development
Radioactive reverberations
kindling for a new world
Employees not allowed
to eat in processing areas
      scrubbing down before they leave
      (some have contaminated bobby pins)
Parker, first Black woman in the U.S.A.
      with a postgraduate
      degree in physics
two masters—the other in mathematics
dedicated, hardworking
Assistant professor
      in physics at Fisk University
close to completing a doctorate at MIT
      afire
microscopic amounts known of her work
within this team constructing secrets:
strikingly more to discover
      from this bold ascending scientist
her time far too short

atomic number 84
leukaemia age 48

✱   *Moments of kindness in a regional hospital*, p. **258**
    The unexpected poetry of PhD acknowledgements, p. **293**

SMOOTH FADE

Zowie Douglas-Kinghorn

In March 2020, the first modern fish extinction was announced to a quiet reception. This was perhaps unsurprising: it occurred during the same month the World Health Organization declared a pandemic, and the smooth handfish, or *Sympterichthys unipennis* had long slipped from view. In fact, the most notable feature of its disappearance seemed to be that it was the first marine animal classified as extinct by the United Nations Red List. Aquatic extinctions have been recorded before, mostly among riverine species such as the Yangtze dolphin, the Mekong giant catfish, India's gharial crocodiles and the European sturgeon. But this was the first time the world acknowledged an absence from its oceans.

Few media outlets included photographs of the singular smooth handfish held as a holotype in archives. Instead, news sites substituted *Sympterichthys unipennis* with images of its more photogenic cousins, including the spotted, red and Ziebell's handfish, all close relations of the smooth handfish but yet to pass over the verge of extinction. Ironically, the stodgy, slippery creature that walks on its 'hands' and has few distinguishing features other than smoothness was one of the first fish species in Australia to be classified in modern taxonomy. French zoologist François Péron collected the first (and only) specimen during an expedition in 1802. It's presumed to have been a common species and captured with a simple dipnet. But it hasn't been seen since.

Tens of millions of years ago, handfish were abundant across the globe, according to the University of Tasmania's Institute for

Marine and Antarctic Science. 'Now, handfish are only found in the waters off southern Australia and Tasmania, where they've earned the dubious honour of being one of the rarest types of fish on the planet,' their website says.

The singular holotype of the smooth handfish remains the only sighting of the species recorded in the past two centuries. Without photographic records showing the animal in its living form, the fish faded without ceremony – that is until the recent declaration of extinction, like a dramatic camera flash illuminating its long absence.

There is a name for the family that the smooth handfish belongs to: cryptic fish. The official name is Brachionichthyidae, which includes the grim-faced anglerfish and monkfish – if you google the latter, it comes up with a recipe for 'Poor Man's Lobster'. The Commonwealth Scientific and Industrial Research Organisation (CSIRO) describes cryptic fish as 'small, camouflaged, or otherwise inconspicuous fish species closely associated with the bottom, which may otherwise be overlooked'.

Looking out at the River Derwent (or as it was known before colonisation, timtumili minanya), its surface appears calm, impassive. I like to imagine there are stones left unturned; that in being overlooked, the smooth handfish remains at large, encrypted in the furling sponge beds and rocky crevices of the estuary. In footage of handfish in captivity, their phlegmy bodies appear translucent, their movements smooth and yet somehow clunky, like a glitch in a video game. Their grumpy, downturned lips and stippled, spiny skins suggest an elderly hermit who has retired from the visible world, an old man retreating into obscurity, sheltered by a wake of billowing sand, bits of information dwindling into speculation.

There are myriad reasons for the demise of cryptic fish. Like many species, the handfish has adapted to survive alongside other organisms, in particular the sea tulip, or *Pyura spinifera*, an ascidian popularly referred to as a sea squirt. It looks a bit like a fleshy maraca. Its tubular stem is where handfish lay their eggs. It may seem unremarkable, but this breeding method sets handfish apart from almost every other fish species in existence. Normally, marine reproduction involves a 'planktonic larval' stage, where masses of newly hatched eggs are dispersed into currents to travel up to hundreds of kilometres away from their original population. Handfish don't do this; their eggs stay put, sequestered in the ascidians until the young emerge as miniatures of their adult parents.

The sea tulip abounds along almost every Australian coastline, but in Tasmania they are particularly vulnerable to predators moving southward on the advancing Eastern Australian Current. One of the sea tulip's fiercest predators is the Northern Pacific seastar. As a flow-on effect, this starfish is a major threat to the handfish's existence. It is effectively a sex pest: it devours the sea tulips, preventing the handfish from reproducing. With the loss of their companion, the symbiotic relationship between the handfish and sea tulip breaks down. These kinds of ripple effects are beautifully encapsulated in Rebecca Giggs' *Fathoms: The World in the Whale*, which illustrates what happens when a co-conspirator in survival is lost: 'So long as such a feature doesn't steer its proprietor into co-extinction, there will be a kind of ghostly residue: a physical communication with no visible respondent. A fascinating notion: that the too-much-ness of the flower might point out a lack in the world.'

The handfish's excesses may seem frivolously indulgent (stepping on webbed fins), even anachronistic (the pearly string of eggs that mould perfectly to the sea tulip's stem). The notion of 'too-much-ness' reminds me of how some academics use the phrase 'more-than-human' to denote the natural world. More-than sounds almost supernatural. The disappearance of handfish will not result in the disappearance of the sea tulip, but its absence will leave a patch of ectoplasm where there once might have been handfish eggs. Similarly, there are the giant kelp forests of southern

Australia, the ancient shellfish encapsulated in middens, the glimmering maireener shells crafted into jewellery by Tasmanian Aboriginal women for millennia.

Ancient organisms trace relationships between people and place, narratives that run deep into geological time.

As their relations continue to disappear from the estuarine coasts, the bereft handfish family appear in images in alley murals, coffee table books and art galleries. In the past, handfish had been observed raising families in beer and rum bottles thrown overboard by sailors. Now, some are stowing their eggs in bespoke pottery; the Hobart-based ceramicist Jane Bamford was recently commissioned in collaboration with the CSIRO to make 3000 artificial spawning habitats from clay to provide breeding platforms for handfish sanctuaries.

One of the handfish's last remaining strongholds is in the D'Entrecasteaux Channel, a shallow estuary about a half-hour drive from Hobart to the south-east of Tasmania, where the ocean's salty mouth intermingles with the freshwater River Derwent. On a map, the channel resembles a key-shaped body of water, locking mainland Tasmania and the smaller Bruny Island apart.

Driving along the Channel Highway through the dormitory suburbs of Snug and Margate, the rugged Alum Cliffs tumble into the Derwent Estuary. The beachside suburb of Electrona is the site of humming salmon pens, shellfish farms and a now-defunct carbide works. Children make sandcastles for marauding soldier crabs along Peggy's Beach. This is where my mother used to clean holiday homes, where the channel's slow, tugging currents are now embroidered with salmon farms.

To my mind, an Atlantic salmon evokes a ballet slipper dancing up waterfalls, surging across oceans in shoals to reach their place of birth. They are the antithesis of the stodgy, graceless handfish, who rarely leave their tidy niche in shallow, salty riverbeds. In Tasmania, the Atlantic salmon is the product of commercial intervention. It is a cold-water species that has been brought to some of the world's fastest warming waters along the state's south-east coast. It is an environmentally intensive process to keep millions of caged fish circling in a column, depositing tonnes of ammoniac and antibiotic-laden waste on the estuary floor. In recent years, pollution from industrial salmon farming has glutted local bays with algae, stripping them of the seagrass beds and sponge reefs where handfish live. Fish farms are a major threat to the remaining numbers of red handfish in the wild, of which there are thought to be just 80 in existence.

Most of the red handfish population is in Norfolk Bay, an inlet along the Tasman Peninsula on Tasmania's south-east coast. In 2018 the state's Environmental Protection Agency permitted a company named Huon Aqua to relocate one of its salmon populations to the same inlet. The reason for moving the fish was their exposure to a fatal infection called pilchard orthomyxovirus. Across the D'Entrecasteaux Channel, salmon farms riddled with bacteria are often subject to mass deaths; over a million fish died suddenly during a marine heatwave at the site in Norfolk Bay where the salmon were previously held.

Neither salmon nor handfish can be removed from their habitat without threatening their existence. In a holotype of the smooth handfish, there is a sense of locality written upon the creature's skin: photographs show flesh that looks like sand dunes where a seagull has made tracks. Other handfish have tiny spinules that stick out like crow's feet, which may have helped against marine invaders. Its smooth cousin did not evolve such defence mechanisms. Certainly, they would be ill-equipped for the influx of chemical pollutants and pest species that pervade their home today. Some might argue smooth handfish had already been on their way out; that these stunted, hand-walking gymnasts were doomed from the beginning.

In 2021, a year after the smooth handfish was declared extinct, a quiet amendment was made to the UN Red List, removing it from the 'Extinct' category and adding it to 'Data Deficient'. Marine scientist and handfish expert Graham Edgar notes that a quarter of the sea-dwelling species on the UN Red List are currently labelled this way, which means they lack the research and information required for credible assessment. Marine extinctions tend to be studied far less than land-based ones, due to the vastness of the ocean and its inaccessibility, as well as a lack of commercial incentive and research funding. 'The pivotal issue associated with assessing the true population status of most marine species, and evaluating the state of the marine environment more generally, is that the marine realm lies out of sight and is expensive to survey,' says Edgar.

To justify its continued existence, an endangered species should be (a) aesthetically appealing or (b) commercially useful. In some cases, such as that of the gastric-brooding frog, whose ability to incubate eggs in its stomach has been of interest to some pharmaceutical companies, revival is lucrative. 'De-extinction' is the word often used to describe the genetic manufacture of dwindling creatures in captivity without necessarily considering how we might restore their habitat to support populations. In the case of the handfish, the potential for using an enzyme in COVID-19 diagnosis was cited by marine scientist Katie Matthews in *Scientific American*: 'It might be hard to imagine why a little organism occupying a small niche in a place few humans ever visit might be important – but it's an enzyme from an extremophile microbe that's being used in tests to diagnose COVID-19 right now. Biodiversity matters, even if you can't see it with your own eyes.'

You can sense the author grappling for utility here, attempting to hook the handfish's survival to current events. It is difficult to justify the ongoing existence of a bottom feeder that has endured millions of years without developing a swim bladder. Contrast this with Edgar's quote from the same article: 'They spend most of their time sitting on the seabed.' Why should anyone care about endangered cryptic fish? The handfish is camera-shy, lazy and commercially useless. And yet in their maladaptive oddness, awkwardness and lack of defences, in their nakedness and smoothness, the same aspects that render handfish vulnerable also make them compelling. Why do they walk on their fins on riverbeds, when their relatives swim at the bottom of the sea with lanterns dangling over their heads, like tiny, water-dwelling miners? Why do they remind us of our own hands?

The marine biologist Neville Barrett notes that in the past two decades, only five ecologists across Tasmania have been employed full-time to study 'unexploited' marine invertebrates. In the absence of funding, hundreds of volunteers each year engage in 'citizen science' to achieve the goals of Australia's handfish recovery plan. These programs involve divers counting fish, monitoring habitats and collecting data alongside organisations like Reef Life Survey, as well as recording sightings by the public. In this way, handfish morph from desiccated flesh into hearsay. These kinds of encounters suggest a literary endeavour, the making of memory and narrative in the absence of empirical research. Incidentally, one of the main supporting foundations for the Reef Life Survey program is also a sponsor of the Emerging Writers' Festival.

One goal of the CSIRO's handfish recovery plan is a public narrative-building exercise to 'promote community awareness of the value of handfishes as part of Australia's unique biodiversity'. Supported projects include visual art and storytelling. In the children's book *Hold On! Saving the Spotted Handfish*, Gina Newton and Rachel Tribout write: 'Have you ever seen a fish that could do a handstand?' The question invites the reader to play with anthropomorphism, as if to say the handfish is just like you, the concept of a bottom-dwelling fish becoming elastic enough to evoke a human body.

The handfish is perhaps more aesthetically pleasing than its closest living relative, the glowering humpback anglerfish, also known as a 'seadevil' or 'viperfish', which inhabits much deeper, murkier territory. Several million years ago, the handfish diverged from its deep-sea ancestors to bypass the need for a swim bladder, the sac of gas that most marine lifeforms need to stay afloat and produce sound.

The handfish lives relatively close to the surface, while most other cryptic fish inhabit a world largely unknown to humans, often referred to as the 'twilight zone'. To quote Shin Tani, the vice-admiral of the General Bathymetric Chart of the Oceans, 'Since 1991, we have known more about the topography of Mars than we do about the earth's seafloor'.

In the introduction to *On Photography*, Susan Sontag writes: 'To collect photographs is to collect the world'. If classifying the visible world involves capturing an animal's photograph, the smooth handfish is a fugitive. It refuses to cooperate in its cultural embalming. Its smooth fade from life has a blurred, obscure quality that forces us to turn to its relatives for evidence. From citizen science to children's storybooks, it is a messy endeavour to mourn such an ambiguous creature, one that resists easy closure in a rapidly changing climate. We have only a holotype, a corpse reworked into a photograph. Even still, the desiccated remains of the smooth handfish bear the centuries-old imprint of the place that cradled its species. Perhaps its fade into the background is a refusal to be mourned. Rather than be integrated into a known, classifiable whole, it becomes an ambiguous loss, an absence that evades definition as its disappearance becomes indistinguishable from that of its home.

Tasmania is famous for the extinction of the thylacine, commonly known as the Tasmanian 'tiger' or 'wolf', despite its closest living relative being the termite-eating numbat. Its populations were not always restricted to Tasmania, and in fact thylacine fossil records can be found across mainland Australia and Papua New Guinea.

Island geographies often crop up in modern-day extinction narratives. Because of their relative isolation, these tiny land masses often preserve idiosyncrasies in the absence of outside threats. According to ornithologist and writer John Woinarski, islands and atolls are where non-threatening plants and animals can evolve in isolation from predators that are more abundant in mainland areas. In the radio series *Extinction Elegies*, Woinarski speaks of 'childlike' and 'naïve' animals, flightless birds and plants that grow without developing thorns, as though they were estranged, sheltered cousins thriving in secluded areas, where there is no need to develop the necessary defence mechanisms of their mainland relations. As a shallow inlet, the D'Entrecasteaux Channel would have been a relatively peaceful haven for species like the handfish to develop gently, tangentially – as holotypes show, many evolved only a few tiny spines, some have no spines at all. In behavioural ecology, such anomalies can be studied to indicate how ecosystems change over time.

Despite being widely marketed as a pristine wilderness, Tasmania bears the mark of heavy industry. In the west of the island, open-cut mining of zinc, copper and silver has led to one of the most severe cases of acid dumping in the southern hemisphere. Over the past century, more than 150 million tonnes of sulfidic acid from the Mt Lyell Mine were dumped into nearby rivers, contaminating waterways with lead, copper, arsenic and cadmium. As a result, the Queen River is famed for its dark orange hue. 'Beneath the King and Queen Rivers and beneath the deep vast waters of Macquarie Harbour lies what is virtually a pavement of copper,' writes Patsy Crawford in *King: The Story of a River*. 'It sits on the floor of every waterway that the effluent from Mt Lyell mine has entered, and is topped up by tailings and acid drainage.' When my family lived in Queenstown, the water from our taps was unsafe to drink.

The remote west coast is also the home of the Maugean skate, a moth-like ray that flits along the edge of extinction in the depths of the Macquarie and Bathurst Harbours. It was first officially named and described in 1988 by Graham Edgar.

There are likely to be other species that perished in the mining waste before being scientifically described.

Karen Gowlett-Holmes, a marine scientist and tour dive operator on the Tasman Peninsula, has described hundreds of species in her lifetime. 'There are still things we don't know about out here,' she says. 'We can probably name around 30 per cent of the species 30 metres below the surface. The other 70 per cent we haven't described. The climate is changing so fast, they're actually becoming extinct before we know that they're there.'

For the estuaries flowing to the west coast from the King and Queen rivers, the situation is grim. 'Only tiny micro-organisms have been found still managing to survive,' writes Crawford. A recent study from the Australian National University found that airborne heavy metals from Queenstown have travelled as far as Dove Lake in Cradle Mountain, contaminating an area over an hour's drive away. Other waterbodies polluted by the Mt Lyell Mine include the Owen Tarn and Basin Lake, where contamination levels of heavy metals are among the highest in the world, according to lead researcher Larissa Schneider, who compared them to highly polluted waterways such as the Kurang River in Pakistan and the Shur River in Iran. Schneider notes more research is needed to determine the impact on aquatic ecosystems, including that of lead levels on deformities in fish: 'There is a case in the US where levels were actually lower than Owen Tarn and Basin Lake and they had serious reproduction problems with the fish there. The levels in Tasmania are even higher.'

Riverine pollution has also played a part in the demise of the handfish in the Derwent. According to a recent *State of the Derwent Estuary* report, 'throughout the 1960s, '70s and early '80s, handfish were frequently seen by divers along the eastern and western shores of the Derwent, and adjoining bays. However, major declines occurred in the mid-1980s and extensive surveys of the estuary floor in 1994 and 1996 found only a handful of specimens at several locations throughout their former range.'

In the mid-1970s, mercury in the Derwent skyrocketed to levels comparable to Minamata Bay in Japan during the early 1900s, when an epidemic of mercury poisoning spread from neurotoxins originating in wastewater dumped from a nearby chemical plant into the Shiranui Sea. In his recent nonfiction book *Toxic*, Richard Flanagan cites large quantities of heavy metals leaching from chemical factories into the water, contaminating fish stocks and resulting in catastrophic health impacts for generations of people in Minamata.

The establishment of the *Environment Protection Act* in 1973 stymied pollution levels in the Derwent estuary. Similar to the west coast's rivers, many of the heavy metals from zinc smelting and pulp milling in the Derwent are now sandwiched between layers of underwater sediment. The cold water suppresses these deposits from leaching into the surrounding ecosystems. But under changing conditions, pollutants can be 'remobilised' from contaminated sediments within the estuary, according to the 2015 report by the State of the Derwent Estuary group. When oxygen levels plummet underwater during high temperatures, heavy metals sequestered underneath the sediment are released through biochemical processes. This often occurs in summer, when nutrient-hungry algae deplete the oxygen levels in seawater – algal blooms, Flanagan points out, are linked to an influx of nutrient-rich wastewater from industrial aquaculture. 'The scale of the salmon farmers' gigantic expansion into Storm Bay could reverse decades of work cleaning up the Derwent by massively increasing nutrient levels,' Flanagan writes.

Similarly to the fish in polluted mountain tarns, handfish act as indicator species for the sea's changing chemistry, a kind of underwater canary in the coalmine.

As I learn more about the handfish, I am reminded of the

phrase 'shifting baseline syndrome', an idea popularised by biologist Dr EJ Milner-Gulland. 'Generational amnesia is when knowledge is not passed down from generation to generation. For example, people may think of as "pristine" wilderness the wild places that they experienced during their childhood, but with every generation this baseline becomes more and more degraded.'

The theory says that the quicker the environment changes, the quicker its preceding state is wiped from memory; the more the biodiversity of an area diminishes, the less the change will be noticed in the future. In this way, collective memory becomes a quicksand, where the shifting baseline becomes the norm for coming generations.

Through their demise, cryptic fish are becoming a viewfinder for a receding baseline. Rising ocean temperatures are limiting the habitat of benthic organisms, which rely on cold, oxygen-rich waters for survival. The advance of the East Australian Current and the Tasman Sea hotspot of ocean warming is interconnected with the fate of myriad other species. The disappearance of the handfish may be quiet, but the constellation of extinctions it heralds is manifold. As Rebecca Giggs points out in *Fathoms*, extinction doesn't happen overnight. An entire species doesn't usually go out with a bang. Nor does this occur in a vacuum. The reality is that many species pass into extinction before they are even noticed by the modern world. It is often a process of slow and silent loss, a smooth fade into non-existence.

In sound engineering, a 'fade' is a gradual increase or decrease in the level of an audio signal, often used to create a smooth transition from one piece of music to another. I am reminded of the words of the landscape gardener and environmental activist Mary Reynolds:

It is so quiet now. Eerily quiet. When you're driving at night your windscreen is no longer covered in dead insects and moths like it used to be when you were a small child. As a species, we immediately forget what is lost and only see what exists right here, right now as the new normal.

The phrase 'the new normal' has been repeated many times in recent years.

As humans, we are designed to adjust to changing conditions. In response to climate change, many creatures will move elsewhere; this includes sea stars and their pesky cousins, sea urchins. But the slow-moving handfish can't disperse its young a long way, or even swim very far on its own. It is not ready for any sudden upheavals.

After two centuries of their home being used as a whaling port, dredged for shellfish harvesting, then as a dump for zinc smelting wastewater and milling effluent, we are now witnessing the disappearance of a creature once abundant enough to be plucked from the water in a colander. Even its disappearance is unsure. Most of what we know about the smooth handfish is inferential. It hasn't been noticed because most of us were never looking – or listening – for it.

Almost half a century ago in 1977, NASA sent two phonographs into outer space, attached to two robotic probes called *Voyager 1* and *Voyager 2*. The records offered discerning aliens a snapshot of life on Earth, containing audio recordings of music by Chuck Berry and JS Bach, the songs of humpback whales and the sound of a human kiss. The astronomer and writer Carl Sagan contributed a recording of his own laughter. He said *Voyager* was a sign of hope that extra terrestrial creatures might be technologically advanced enough to encounter these human cultures from afar.

It seems a bit of a stretch to expect intergalactic life forms, however technically advanced, to be vinyl enthusiasts. The ostensible aim of the Voyager Golden Record, to be 'heard' by bringing snippets of human life to aliens, seems unlikely; more realistic is the possibility of the craft existing indefinitely, suggesting the possibility of being heard. The science communicator Ann Druyan said the point was to 'send a piece of music into the distant future and distant time, and to give it immortality'. Sagan likened the Golden Record to a message in a bottle thrown into the ocean, which is an interesting metaphor – not least because the earth's climatic regulatory systems, ocean currents, are rapidly breaking down.

Our oceans currently absorb 80 per cent of the atmospheric carbon dioxide and the heating of the planet. Marine temperatures are rising at a faster rate than land temperatures; Arctic and Antarctic ice is melting so fast that many ocean currents crucial to the temperature regulation of the planet are slowing to a halt. Combined with the effect of deoxygenation and acidifying sea pH levels, many underwater ecosystems are on the verge of collapse. In fact, some scientists predict that Earth could reach Permian-Triassic levels of marine extinction by 2300 if greenhouse emissions continue along their current trajectory.

This is a sobering prospect. The Permian extinction is Earth's most severe extinction event to date. It took place 252 million years ago, when huge swathes of the planet became uninhabitable due to subsea volcanic eruptions that added between 3900 and 12 000 gigatonnes of carbon dioxide to the ocean–atmosphere system, increasing atmospheric carbon dioxide levels from 400 parts per million to 2500, a shift that would wipe out 81 per cent of marine species and 70 per cent of land animals. The impact of the asteroid that wiped out the dinosaurs 187 million years later looks puny in comparison.

In the year the Golden Record was launched into space, carbon dioxide composed 333 parts per million of the atmosphere. In the year I was born, 1997, this number was 363. In the year 2023, carbon dioxide makes up 421 parts per million. While the Golden Record spins though space, noise pollution ripples across the earth's oceans.

The whale songs that were specific to the Golden Record no longer exist as mating calls; the silence necessary for underwater communication has been obliterated. The world is increasingly animated with the sound and movement of lifeless engines, while the sounds of living organisms become quieter. Scientists predict that 40 per cent of the fish currently living in the twilight zone – the home of the handfish's deep-sea-dwelling relatives – could be gone by the end of this century alone.

We are living in a twilight zone of extinctions, where ancient organisms are rapidly disappearing. I am reminded of Joan Didion's *Blue Nights*, a meditation on loss and ageing: 'Fade as the blue nights fade ... Go back into the blue ... The fear is not for what is lost ... The fear is for what is still to be lost. You may see nothing still to be lost.'

The fading away of countless species brings to mind a quieter, lonelier world.

I think of the Great Silence theory, or silentium universi, inspired by physicist Enrico Fermi. The paradox asks why humans haven't come across alien life, given the vastness and multiplicity of stars and planets with similar positioning to Earth and its sun. Some people (UFO truthers) say that we have. Others say the universe is simply so big that we can't have come across them yet. But then: 'The Fermi Paradox is a very large extrapolation from a very local observation. You might just as well look out your window and conclude that bears, as a species, couldn't possibly exist because you don't see any,' says the website for the Search for Extraterrestrial Intelligence (or SETI) Institute, a NASA-affiliated organisation, similarly to the orchestrators of the *Voyager* probes, the Carl Sagan Centre for Research.

There is a multitude of life that remains obscure to us on Earth. Some of this life belongs to cryptic fish who walk on their hands and lay their eggs in rum bottles and sea tulips. Tiny lives used to populate our world in droves: the Christmas beetles adorning the pavement, moths barbecuing themselves on car grilles. Their numbers are plummeting, fading prematurely. Their images are not printed on pages or archived in museums. There are many more like *Sympterichthys unipennis*, who disappear not because of asteroids, but at the hands of humans.

❋ *Cat-astrophe: Australia's feral cat problem*, p. **33**
     *A museum heist 70 years ago is still causing a flutter in
        butterfly science today*, p. **275**

WHY CAN'T WE REMEMBER OUR LIVES AS BABIES OR TODDLERS?

Donna Lu

Life must be great as a baby: to be fed and clothed and carried places in soft pouches, to be waved and smiled at by adoring strangers, to have the temerity to scream because food hasn't arrived quickly enough, and then to throw it on the ground when it is displeasing.

It's a shame none of us recalls exactly how good we once had it.

At Christmas, I watched my daughter, somehow already a toddler, being passed between her grandfathers and thought, wistfully: she won't remember any of this. In parks, I push her endlessly on swings, making small talk with fellow parents who have been yoked into Sisyphean servitude, and think, ruefully: why won't she remember any of this?

In 1905, Sigmund Freud coined the term 'infantile amnesia', referring to 'the peculiar amnesia which, in the case of most people, though by no means all, hides the earliest beginnings of their childhood'. More than a century later, psychologists are still intrigued by why we can't remember our earliest experiences.

'Most adults do not have memories before two to three years of age,' says Professor Qi Wang at Cornell University. Up until about age seven, memories of childhood are typically patchy.

Until relatively recently, researchers thought that young brains weren't developed enough to form lasting memories. But studies in the 1980s showed that toddlers as young as two can form memories and recall events from months earlier in great detail. Exposure to early childhood trauma is also well documented to increase the risk of later anxiety and depression. The paradox of infantile amnesia, says Cristina Alberini, a professor of neural science at New York University, is 'How is it that those experiences affect our life forever if they are forgotten?'

Alberini's research in animals has found that memories formed during the infantile amnesia period are, in fact, stored in the brain until adulthood, even though they aren't consciously remembered. In both animal and human adults, forming and storing long-term memories about one's life experiences isn't possible without a region of the brain known as the hippocampus.

Alberini's work has shown that the region is also important in early memories and suggests that infantile amnesia occurs because of a critical period where the hippocampus develops due to new experiences. 'It makes a lot of sense with all the literature of trauma,' she says. 'If the children are learning difficult situations in early childhood, maybe they don't remember the specifics, but their brains are going to be shaped according to that experience.'

Why Māori memories emerge earlier

Differing experiences may also explain why the age at which people recall their first memories varies significantly. Wang, an expert in how culture affects autobiographical memory, has shown that the earliest memories in Americans date from an age of about 3.5 years, almost six months younger than in Chinese people. The American memories tended to be more self-focused and emotionally elaborate, while the Chinese recollections tended to centre on collective activities and general routines, she found.

'In the Asian context, identity and sense of self is less defined by being unique, but [more] about your roles and your relationship with others,' Wang says. To that end, memories may be less important for defining identity than for informing behaviour and imparting lessons. 'If you want to use memory to construct a unique sense of identity, you probably remember a lot of idiosyncratic details,' Wang says.

Another explanation for the discrepancy seems to be how parents discuss past experiences with their children. In New Zealand Māori, first memories emerge earlier than in those of a European background, at about 2.5 years old. Professor Elaine Reese at the University of Otago, who studies autobiographical memory in children and adolescents, points to a strong emphasis on oral traditions in Māori culture but also elaborative conversations when reminiscing about past events.

Reese has tracked groups of children from toddlerhood to adolescence, finding that individuals who had richer narrative environments in childhood could recall earlier and more detailed first memories as teenagers. This was the case for children whose mothers asked open-ended questions and were more detailed when talking about shared past experiences, as well as children who grew up in extended family households.

'We know that from the time [children] are, say, six-month-old babies, they're capable of some kind of mental imagery of something that happened from the previous day or week,' Reese says.

'It's taking that mental image and describing it in words that I think is so important for helping them to hold on to that memory over a lifetime.'

Ironically, for parenting influencers who post about elaborate holidays in the name of creating 'core memories', the early events that children retain can be surprisingly mundane – 'things that most parents would never reminisce elaboratively about', Reese says. 'The classic example from my own research is a child who remembers seeing a worm on the footpath one time.'

There is debate between memory experts as to the role of language in infantile amnesia. Human researchers suggest memories may be limited by an inability to give language to early experiences. 'But there must be something more fundamental that also plays a role because we see this same [infantile amnesia] effect in non-linguistic animals like rats,' says Professor Rick Richardson of the University of New South Wales.

'Improbably early'

The brain lays down memories not as discrete files as on a computer but as networks of neurons across the brain. Recalling a memory activates those networks and strengthens the links between neurons. This is not to say memory is stable: 'Every time you revisit a memory and think about it, you're changing it,' Reese says.

Repeated suggestions can lead people to create images and form false memories, Wang says, citing a famous case in Jean Piaget, the influential child development psychologist. Piaget had a clear memory of his nanny fighting off a would-be kidnapper when he was two – but years later, she confessed that she had fabricated the story.

In a 2018 survey, 39 per cent of respondents reported their first memories occurred at age two or younger. The researchers suggested that 'improbably early' memories, such as recollections of being pushed in a pram or walking for the first time, were likely fictional and based on photographs or family stories. But though memory is malleable and young children are more suggestible, 'confabulation is not that common', Wang says. 'Under normal conditions, even children do not just take for granted whatever you tell them and incorporate those memories.'

So if experiences of our early milestones – first birthday, first steps, first trip to the beach – seem to be cached somewhere in the brain, why can't we consciously access them? While psychologists say it can be adaptive to forget, that doesn't explain why the memories formed before age seven seem to decay faster than when we're adults. Alberini hypothesises that early unrecalled memories may function as schemas upon which adult memories are built. Like the foundations of a home, they remain concealed but crucial.

✱ *The night I accidentally became a corpse flower's bedside manservant*, p. **9**
Some psychedelic medicine developers want to ditch the therapy aspect. What could go wrong?, p. **120**

USING TRASH TO TRACK OTHER TRASH

Clare Watson

Australia's vast coastline is littered with marine debris. From burst balloons and countless straws to plastic drink bottles, styrofoam and fishing lines, all sorts of trash ends up on the country's beaches, and Heidi Tait, co-founder of the nonprofit Tangaroa Blue, has combed through it all. But as the old adage says, some trash is actually treasure provided you look at it from the right perspective. In this case, Tait and the Tangaroa Blue volunteers working to clean up Australia's beaches unexpectedly accumulated a trove of strange tyre-shaped capsules scattered along the Cape York coast, near Australia's north-eastern tip.

When Tait and her teammates started finding the capsules washed ashore, they weren't quite sure what they were looking at. But by busting one open, looking at the company names listed inside, and making a few calls, Tait eventually connected with Satlink – a Spanish satellite communications company. Satlink's GPS-enabled buoys, the ones the beach cleaners kept finding, help commercial fishers track their nets, lines, and other gear.

Tait's partner, Brett Tait, Tangaroa Blue's circular economy developer, had a brainwave that would see the buoys not just recycled but reused.

For more than a decade, boat crews working farther west, in Australia's Gulf of Carpentaria, had been telling the Taits about how abandoned fishing nets were circling the gulf, ensnaring and strangling sea turtles and dugong. These so-called ghost nets had either broken free from commercial fishing vessels and gotten lost, or were cut loose by fishers after getting snagged on rocks. Weighing a few tonnes each, the nets that boat crews had chanced upon in the gulf were often too big for them to heave out of the water. They'd typically report the finds to the authorities, but by the time anyone with an appropriately equipped vessel could head out to retrieve one, the mass of tangled rope had often vanished from sight.

Perhaps, Brett thought, Tangaroa Blue could solve the problem using their newfound GPS buoys. 'The trackers are such a high-tech piece of equipment,' Tait says. They're obviously not cheap, and for them to go to a landfill 'seemed like such a waste'.

So, in December 2022, Tangaroa Blue started handing the buoys out to its flotilla of local partner crews: boat charter operators, commercial fishers, Indigenous rangers, and National Park Service members who had agreed to carry the repurposed buoys. When these teams come across a ghost net they can't haul in, they hook one of the GPS-enabled floats onto it. Once attached, the tracker pings its location every few hours. A web portal lets Tangaroa Blue monitor the nets' movements and alerts the organisation if one is drifting dangerously close to a coral reef or shipping lane.

But finding and tagging the nets is only the first step; coordinating their recovery is the real challenge. The trackers can measure the weight of the rope dangling beneath the float, so Tangaroa Blue's staff know how big a boat they need to charter to retrieve it. But 'we don't have an unending supply of money to hire vessels', says Tait.

So far, crews from around 100 commercial fishing vessels representing 22 international companies have gotten on board with Tangaroa Blue's project; they've given Satlink permission to reassign any of their lost buoys that Tangaroa Blue finds for the nonprofit's use. Eventually, Tait hopes to encourage fishing companies to chip in to a recovery fund that would pay for charter boats to retrieve nets without delay.

The repurposed GPS trackers have so far helped Tangaroa Blue recover three ghost nets from Australian waters – including one particularly pesky net that had been spotted in the Gulf of Carpentaria multiple times by local boaters and evaded recovery for about a year. The net weighed three tonnes, and just three weeks after being tagged, it was hauled in on the gulf's eastern flank, near the town of Weipa.

Tangaroa Blue's idea to tag and track ghost nets isn't entirely new; researchers with the Commonwealth Scientific and Industrial Research Organisation (CSIRO), Australia's national science agency, suggested it a decade ago. Taking the idea on board, Australian maritime authorities tried using battery-powered tags to monitor a few ghost nets. Tangaroa Blue's approach differs in that its recycled buoys are solar powered and require less upkeep. The nonprofit's assembly of volunteer ghost-net spotters also greatly expands its reach.

From an environmental perspective, the GPS-fitted floats themselves are a massive plastic pollution problem. Researchers estimate that commercial fishers deploy between 46 000 and 65 000 GPS trackers in the Pacific Ocean each year. When the devices are lost or discarded, ocean currents and prevailing winds push many of them around Papua New Guinea and toward Australia's northern coast. The Gulf of Carpentaria, in particular, is a magnet for both ghost gear and lost buoys. Its two long outstretched arms, shallow water, and clockwise gyre all work together to create a vortex that drags in junk fishing gear from all over.

In 2013, Denise Hardesty, a CSIRO marine scientist who had been tracking ghost net numbers in Australian waters, identified a narrow area where ghost gear seems to get sucked into the gulf. If Tangaroa Blue focuses efforts on that spot, her work suggests, it could intercept some nets before they ever get a chance to impact the wildlife that dwells in the gulf's shallow water.

As supportive as he is of Tangaroa Blue's approach, Erik van Sebille, an oceanographer at Utrecht University in the Netherlands, says there may be limits to the technique's transferability to other countries similarly swamped by ghost gear. Many Pacific Island nations, for example, might lack access to big vessels that could recover tagged ghost nets quickly. Ultimately, he adds, rather than looking for new ways to clean up lost nets, 'it would be much better to prevent ghost nets from being in the ocean in the first place'.

✱   *Solomon Islands tribes sell carbon credits, not their trees*, p. **26**
     Fishing for a glacier's secrets, p. **58**

ALTERNATIVE ACCOMMODATION

Anne Casey

It wasn't a perfect fit, but these days
with the pressure on housing stock,
everyone was having to make do
and he quite liked the colour,
though he knew a few who
might have been put off
by the coral-red walls
with matching cabinetry
and the tessellated ceiling.

He hadn't expected the spare room
and it had come fully furnished—
though the built-in bed had an awkward
dip in the middle, there were other
creature comforts well exceeding
his humble expectations and
superior in every respect
to his prior lodgings whose walls
had given off a faint fishy whiff
suggestive of previous tenants'
dietary preferences (or so
he hoped as he shelved
suspicions
regarding their
hygiene habits).

As a bachelor, with admittedly
limited ambition, and little schooled
in domesticity, a spare antechamber
with space enough to swing a catfish sprat
was an unanticipated indulgence.
And who could quibble with
the expansive sea view
from his very own
salmon-tinted
dining-room?

Author's note: In January 2024, the BBC reported scientists had discovered two-thirds of hermit crab species worldwide were living in 'artificial shells' using items discarded by humans, including toy plastic building blocks, due to a shortage of natural shells and inundations of ocean-borne pollution. It is estimated there are 171 trillion pieces of plastic in the world's oceans.

✱ *'Earth poetry' in the Arctic*, p. **53**
   Smooth fade, p. **188**

ETHICAL PROBLEMS CONTINUE TO PLAGUE BIOMETRIC STUDIES OF CHINESE MINORITY GROUPS

Dyani Lewis

Yves Moreau thought something was amiss when he came across a genetics paper about Tibetans in China. In the 2022 report in *PLoS ONE*, a team of researchers had collected blood samples from hundreds of people in the Tibet Autonomous Region of China and recorded genetic markers on their X chromosomes. The researchers concluded that this analysis was useful for forensic identification and paternity testing.

The paper raised immediate red flags for Moreau. Over the past half-decade Moreau, who is a computational geneticist at the Catholic University of Leuven in Belgium, has become deeply concerned about the ethics of studies that report the collection of biometric data from vulnerable or oppressed groups of people.

In this case, he worried that Chinese security forces might have been involved in the work. One concern was that the blood was collected by being blotted onto reference cards – a method of choice for police forces. Moreover, in 2022, the international advocacy organisation Human Rights Watch, among others, had reported that a mass DNA-collection program of Tibetan populations was underway. Moreau also recognised one co-author from other papers he had flagged: Atif Adnan, who had previously been based in China and was now affiliated with the Naif Arab

University for Security Sciences in Riyadh. For Moreau, this raised questions about links to security forces. Moreau urged the journal editors to investigate whether the Tibetans in the study had given informed consent.

And by January 2023, three months after Moreau's complaint, the publisher PLOS, based in San Francisco, California, had retracted the article, with a notice saying that the editors had concerns about informed consent and ethics approval procedures that were not resolved by documents they'd been sent.

But the alacrity of this retraction is unusual. Moreau and a few other researchers have alerted publishers to 96 papers over the past half-decade, and raised questions about genetic databases that hold data from minority ethnic groups. Ethical concerns are particularly acute in forensic science because the field has close connections with law enforcement, Moreau notes. So far, however, only 12 of the 96 flagged papers have been retracted. In most cases, decisions on whether to retract a paper are still pending – some more than three years after Moreau raised his concerns. He adds that he has found hundreds more articles that he has yet to challenge. Journal editors say that investigations can be lengthy because they are complex. But Moreau says that 'the inordinate delays by many publishers in issuing decisions amount to editorial misconduct'.

Origins of a quest

Moreau first became alerted to the ethics of widespread DNA profiling in 2016. That year, he learnt of a state-run program in Kuwait that was set up by law to collect and catalogue genetic profiles from its citizens as well as visitors to the country. Moreau urged the European Society of Human Genetics (ESHG), a non-profit organisation based in Vienna, to take a public stance against the move, which it did in September 2016. After media attention, the Kuwaiti parliament repealed the law a year later. The success of the campaign was intoxicating. 'I felt like the lucky guy who goes to the casino and bets everything on seven and wins,' says Moreau. 'Of course, I had to do more.'

The opportunity came quickly. In 2016, Moreau learnt that DNA profiling was being deployed as part of the passport registration process in China's north-western province of Xinjiang. The region is home to the mostly Muslim Uyghur minority ethnic group, which has been the target of surveillance and mass detentions condemned by the international community. (China's government says that its operations in Xinjiang are aimed at quelling terrorist activities.)

Moreau contacted the China arm of Human Rights Watch to offer his expertise. He also searched the academic literature and found dozens of papers describing the genetic profiling of Uyghurs and other minority ethnic groups in China. Other papers described research to distinguish people's ethnicity from their faces: journalists have since reported that authorities in Xinjiang have used surveillance cameras with facial recognition software to identify Uyghur faces and that, in government contracts, Chinese firms that develop facial recognition software and cameras say they offer the ability to recognise the faces of Uyghur or Tibetan people.

Moreau says that such papers should be retracted not only because it can't be guaranteed that the people involved truly gave free informed consent – given the societal conditions at the time – but also because journals should deny researchers that are doing this work the credit of internationally published academic articles. When the collection of biometric data, including DNA and facial scans, is part of a system of oppression, scientific publishers should act so as not to be complicit in such systems, he says. Moreau notes that although journal papers and databases have also analysed data collected from other oppressed groups potentially without their consent (such as the Roma), he has concentrated mostly on flagging studies from China, in part because of China's widely documented large-scale DNA-collection efforts, and because work on Uyghurs and other minority ethnic groups is heavily over-represented in Chinese forensic genetic papers.

When asked about concerns over the use of biometric data in China relating to minority ethnic groups, a Chinese government representative told *Nature*: 'China is a country governed by law. The privacy of all Chinese citizens, regardless of their ethnic backgrounds, are protected by law.'

Informed consent

The idea that signed consent forms don't always prove that someone has given voluntary, informed consent is borne out by reports coming out of Xinjiang. Abduweli Ayup, a Uyghur linguist who was detained and imprisoned in Xinjiang, attests to the circumstances that lead people to provide blood, saliva or urine samples under duress. Ayup, who has written a book about his experiences, related them to *Nature*.

In August 2013, police arrested Ayup and took him to a detention centre in Kashgar, in western Xinjiang. Chinese authorities initially told him he was being detained for his involvement in a Uyghur separatist movement, but then they charged him with economic fraud; Ayup says none of the accusations are correct. During his 15-month detention, Ayup says, he was subjected to inhumane treatment by physicians and nurses at the centre. Prison officers often forced him to take unidentified tablets, he says, and then he and other detainees would be told to strip naked and line up so that nurses could administer a health survey and collect blood samples. 'I felt like a mouse' being experimented on, says Ayup, who is now based in Norway but continues to worry about his brother and sister, who, as far as he knows, remain in detention.

Ayup says that no consent forms obtained in Xinjiang can be trusted, especially those gathered since around 2014, when the Chinese government began clamping down on Uyghurs and other minority ethnic groups in the region. 'No one says, "no",' says Ayup, even outside prison, because people are afraid of being arrested. During detention, Ayup signed consent letters to say that he was freely participating in the process of blood sample collection. But opting out was not an option. 'How don't we sign it?' he asks. 'We are prisoners.'

Publisher actions

For the 12 papers retracted so far in relation to this issue, publishers say they've done so on the grounds that they haven't been able to establish that participants gave informed consent. (Publishers have also closed seven cases deciding no action was warranted.) But around 70 papers are still under investigation two or more years since Moreau first flagged them. That includes 14 papers in the journal *Molecular Genetics and Genomic Medicine*; in 2021, nine members of the journal's editorial board resigned in response to the journal's failure to tackle the concerns raised by Moreau. Wiley, the journal's publisher based in Hoboken, New Jersey, says that it is still investigating.

For Moreau, the involvement of law enforcement in a study is a clear sign that a journal should investigate. Of the 96 papers that Moreau or others have flagged to publishers, 60 per cent have at least one co-author who works for a public security bureau or other law enforcement entity. Other papers enlist police officers in sample collection, which also calls into question whether consent was freely given, says Moreau.

Dennis McNevin, a forensic geneticist at the University of Technology Sydney, who co-authored another paper that Moreau has queried, says that in many countries 'it is not unusual for police to help facilitate forensic population-genetics research'. McNevin's work, published in 2018 in *Scientific Reports*, related to analysis of DNA from 1842 people from four ethnic groups in Xinjiang. The paper included individual (anonymised) genotype data in its supplementary information. Following Moreau's request in 2022, the London-headquartered publisher Springer Nature issued a correction that removed the genotype data. (*Nature*'s news team is editorially independent of its publisher Springer Nature.)

Moreau has other concerns about the paper. He filed a freedom-of-information request to the University of Canberra – where McNevin worked at the time of publication – to find out more about how the data were collected. McNevin probed further and was told in an email from co-author Adnan (also a co-author of the retracted *PLoS ONE* article), a forensic geneticist then at China Medical University in Shenyang, that a police officer was present during sample collection and consenting procedures. Adnan also said that DNA donors would sometimes give a thumbprint, with a local guide – a physician or police officer – signing the consent form on their behalf.

However, in response to *Nature*'s inquiries, Adnan said that no participants in this study consented with a thumbprint, and no police officers were involved in the collection of samples. 'I am sure that there wasn't any involvement of any law enforcement agency at any point during this or any research project,' he said. He adds that he doesn't know why he told McNevin that a police offer was involved (see the update at the end of this article).

McNevin says he has no evidence that participants were coerced to take part, and that, as far as he knows, police were not involved in data analysis and do not have access to the data. (He also shared emails showing that the genotype data that the publisher later removed in its correction were first added as a supplementary file at the request of the journal's editor, and were not included in the original submitted manuscript.) But Moreau says that 'the simple presence of a police officer in a community where mass persecutions are ongoing is enough to void the validity of any consent'. Moreau relayed his concerns about police involvement to Springer Nature in 2022 as well, but they were not followed up until *Nature*'s news team contacted the publisher for this article. Tim Kersjes, head of

Research Integrity, Resolutions at Springer Nature, says that the publisher will investigate the concerns further.

In some instances, publishers have closed their investigations. The journal *Genes*, for instance, decided that no action was needed for seven articles that it published. MDPI, the journal's publisher in Basel, Switzerland, says that authors sent ethical oversight documentation, and the studies' institutional review boards confirmed their validity. In one article, the authors report investigating the genetic origins of the Hui people, a Muslim minority ethnic group mainly in northern China. Several of the authors work for the Academy of Forensic Science in Shanghai, part of China's Ministry of Justice. They collected blood samples and demographic information from 98 Hui individuals, and Moreau questions whether consent was freely given. In another paper, authors were affiliated with the Criminal Investigation Department of Yunnan province and the Public Security Bureau of Zibo City in China. (The authors of these papers did not respond to inquiries for this article.)

Moreau disputed the decisions to retain the *Genes* papers with MDPI and the Committee on Publication Ethics (COPE), an industry body based in Eastleigh, UK, that guides publishers on best practice. But Iratxe Puebla, COPE's facilitation and integrity officer, told *Nature* that COPE only reviews how a journal follows up on concerns, and not the scholarly content of the articles or specific editorial decisions.

Updating consent policies

In the wake of Moreau's complaints, some publishers have updated their policies on informed consent, and their guidance to editors on considering work from potentially vulnerable groups, generally to be clear to authors that extra scrutiny might be required. Springer Nature, for instance, updated its policies at the end of 2019; it now requires that editors take extra care with studies involving vulnerable groups because of the risk of coercion, and notes that authors must supply documentary evidence of consent when requested. And although MDPI has not retracted or corrected any papers, it has also updated its processes. Since mid-2021, its policies have stated that editors will subject studies that involve vulnerable groups to extra scrutiny, and could request further documents from the ethics boards involved.

The Institute of Electrical and Electronics Engineers (IEEE) in New York City had no policies around informed consent until it updated them in September 2020 to require authors to confirm that they had participants' consent and approval from an institutional review board. But the publisher decided not to apply this policy retroactively to retract two articles flagged by Moreau that were published in 2010 and 2017. The former describes a database of facial images from three minority ethnic groups to develop an algorithm that can determine ethnicity; the latter, the creation of a database of facial images from people of various ethnicities in Xinjiang. Instead, IEEE issued expressions of concern stating that the publisher didn't have a policy of requiring informed consent at the time, and that it cannot now confirm whether participants gave informed consent. A third paper, also flagged by Moreau and published in 2019, was retracted: not because of concerns over informed consent, but because the data underlying the study were not accessible. However, the retraction notice also says that IEEE was unable to confirm whether consent was obtained from a person whose image is shown in the paper. The authors of these papers did not respond to *Nature*'s inquiries.

Virginia Barbour, editor-in-chief of the *Medical Journal of Australia*, and a former chair of COPE, is critical of journals' inconsistencies and lack of transparency when their articles are under investigation, especially when the investigations become protracted. After two or three years without resolution, she says, a paper should still be updated to alert readers of an ongoing investigation. Of the 74 articles still under investigation by publishers, 47 do not alert readers that an investigation is going on.

In some cases, publishers have made it clear that the receipt of informed consent forms is their key focus. One example is an investigation by Wiley into a study that trained algorithms to distinguish the faces of Uyghur people from those of Korean or Tibetan ethnicity. In 2019, Wiley told Moreau and others that it had concluded from consent forms and university approval documents that the authors had gained consent from the students at Dalian Minzu University in China whose faces were used in the research. 'We are aware of the persecution of the Uyghur communities,' Wiley said. 'However, this article is about a specific technology, and not an application of that technology.'

The issue was further complicated in 2021, when Curtin University in Perth, Australia, asked for the study to be retracted; one of the co-authors, Wanquan Liu, had an affiliation with Curtin, but the university said it was not aware of the work and had not given him ethical approval. Instead, Wiley removed the Curtin University affiliation in January 2022. Liu did not respond to *Nature*'s inquiries.

Moreau says that he has since provided Wiley with further concerns, in the form of a masters thesis that suggests that data in the paper were not collected as described in 2014, but were collected at least two years earlier, which calls into question the authors' statements to Wiley about when consent was obtained. A spokesperson says that Wiley plans to make a decision on the article shortly.

Some editors are questioning the veracity of signed consent forms even when they are provided. In 2021, concerned by reports of human rights abuses in China, David Curtis, a human genetics researcher at University College London, resigned as editor-in-chief of the journal *Annals of Human Genetics*, published by Wiley. Curtis says he felt he could no longer impartially consider submissions to the journal from researchers in China because of his concerns over whether informed consent claims in such papers could be trusted.

Journals' responsibilities

Henryk Szadziewski, an ethnographer and director of research at the Uyghur Human Rights Project (UHRP), a Washington DC–based advocacy organisation, says that journals need to further tighten their procedures.

Szadziewski was involved in alerting editors to problems with a *BMC Public Health* study published in April 2023. It was conducted on 'Uyghur residents' in areas of Tumshuq, a city that is controlled by the Xinjiang Production and Construction Corps (XPCC), which was under sanction in the United States, the European Union, the United Kingdom and Canada for human rights abuses against the resident Uyghur population at the time the article was submitted for publication in October 2022. Three of the article's authors were affiliated with the Shihezi University School of Medicine in Xinjiang, which Szadziewski says is run by the XPCC; an ethics committee at the university approved the study.

After Szadziewski alerted the journal, the paper was swiftly retracted in August 2023, because the authors 'had not obtained appropriate ethical approval' before recruiting people for the study, the retraction notice said. Szadziewski suggests checking author affiliations for sanctioned organisations (although this would not have flagged XPCC's name, which appears only in the paper's methods section as a source of participants). Chris Graf, research integrity director at Springer Nature, which publishes *BMC Public Health*, says that the publisher already complies with sanctions requirements and didn't breach them in this case, but adds that 'we do not consider ethics to be a box-ticking exercise and we do not consider this the end of the matter'.

DNA databases

Some researchers say the concerns go beyond scientific publications. Data collected during DNA profiling studies is commonly deposited into genetic databases, which are resources for medical researchers, population geneticists and, in some cases, law-enforcement agencies. In 2021, *Nature* reported the concerns of geneticists about the contents of the Y-chromosome Haplotype Reference Database (YHRD), a public repository of genetic markers on Y-chromosomes from men across the world, which shows how these markers are related to male lineages in more than 1400 populations. Police can use it, for instance, to quickly help calculate the likelihood that markers from crime scene DNA match those of a male suspect. Researchers have uploaded anonymous profiles from almost 350 000 males. The database holds samples from Uyghur, Roma and other oppressed minority populations that, Moreau and others have argued, could have been obtained without informed consent.

YHRD curator Sascha Willuweit, a forensic DNA specialist who works for the Berlin government, says that when a research paper is retracted, uploaded profiles originating from that paper are removed from the platform. And in 2022, he told *Nature* that in that year, the YHRD had retroactively requested information on the informed consent and ethical approvals processes for all data uploaded that weren't related to a peer-reviewed paper.

In October 2023, the Charité research hospital in Berlin – which until the end of 2022 hosted the YHRD – informed Moreau by email that data from one retracted study, of almost 38 000 genetic profiles of men from 70 populations in China, had been removed from the YHRD. However, as only data collected as part of that study were removed, thousands of profiles that the study described, which had been obtained for previous papers, were still accessible. Twenty-three of those previous papers were published in the journal *Forensic Science International: Genetics*, and are still under investigation by its publisher Elsevier, based in Amsterdam.

Researchers agree that it's important for the YHRD to represent all populations, to reduce bias when the database is used to analyse DNA collected at crime scenes. But they differ on whether – and on what basis – data from oppressed minority groups should be removed from the database. Maria De Ungria, a population geneticist at the Philippine Genome Center in Quezon City, says that a key question is whether Uyghurs or other communities want their DNA to be part of such a database. If not, 'then of course, we have to respect that's their choice. It's their DNA', she says. And Szadziewski says that the UHRP is preparing a statement that Uyghurs are not willing participants in DNA profiling studies. But Martin Zieger, a forensic geneticist at the University of Bern, says that in his view, a good and broadly applicable justification for removing data should be when a paper is retracted for lack of consent, rather than a community statement.

In 2022, the International Society for Forensic Genetics based in Mainz, Germany, established a Forensic Databases Advisory Board (FDAB) to make recommendations on best practice for the YHRD and other genetic databases used by the forensic science community. In February 2023, the FDAB released its first report, which suggested that database curators should undertake a case-by-case assessment of data sets, discarding those for which there is a high chance that no informed consent was obtained, unless the data were collected before the 1997 UNESCO Universal Declaration on the Human Genome and Human Rights.

Moreau takes the wider view that any broad indiscriminate collection of DNA by authorities is harmful, and thus so are any forensic databases built from such efforts. 'They are part of that structure of social control that terrorises a population,' he says. Noting China's mass collection of DNA from men across the country, as well as in Xinjiang, he would like to see all data obtained in such a way removed from international databases, including data on people from the majority Han Chinese ethnic group.

More papers to come?

Moreau says that there are more worrying papers in the literature that he does not have time to chase down. In March 2021, he found 305 articles published between 2019 and 2021 on forensic genetic studies of Chinese populations in a Web of Science search. Half of the papers involved the police or a judicial authority, and 30 per cent (92 studies) were on Tibetan or Muslim minority ethnic groups.

Moreau now estimates the number of Chinese forensic genetic studies – published before 2019 and since that search – could exceed 500, although it seems that the number of concerning papers has slowed in the past three years. And because of the entanglement of forensic population genetics research and law enforcement in China, 'all such studies should be retracted', he argues.

Moreau says his campaign is also about researchers earning the trust of the broader community, especially that of vulnerable groups. Last November, he won an award for promoting quality in research, from the Einstein Foundation Berlin, for his work advocating ethical standards in the collection of human DNA data. 'Once people understand that there are people in the [scientific] community that are willing to champion or to challenge the status quo, they actually will feel safer,' he says.

Author's note: After publication of this article, Atif Adnan contacted *Nature* with more details about his role in a project to analyse the DNA of people from different ethnic groups in Xinjiang. He says that his role in the project was in 'ensuring the accuracy and integrity of our research findings' and that 'the complexities surrounding the ethical and political aspects of sample collection, particularly in international contexts, fall outside my purview'.

✱ *China's race for fusion energy*, p. **44**
Gaza: Why is it so hard to establish the death toll?, p. **169**

FROM HYPNOTISED TO HERETIC: IMMUNISING SOCIETY AGAINST MISINFORMATION

Linda McIver

I want to talk today about feelings, because I don't think we talk about them enough in the tech industry. I've learned a lot from my cat, who has big feelings about many things, including her preferred human leaving for work, the urgent need for snacks, her desperate lack of playmates at 3 am, and people who DARE to move without permission. When she has big feelings, she likes to tell us about them. At length. And I am currently having big feelings about the tech industry, so I'd like to tell you about them.

The tech industry sells us 'smart' watches that can't actually reliably do any of the things you buy them for. The most positive study of sleep trackers says they're 60–75 per cent accurate at picking the sleep stage AT BEST. In the case of my Garmin Vivomove, it has repeatedly claimed I was in REM sleep while having breakfast, making a cup of miso soup, and putting a load of washing on. Which is remarkably productive for someone who is allegedly still asleep.

My husband's Fitbit recently recorded him having a swim – only a short one, 208 metres – while he was folding washing. I like to imagine a series of parallel worlds where other versions of us are doing the things our 'smart' watches say we are.

What's more, my watch can't even reliably tell the time! Periodically the hands of my watch just get lost and tell some completely other time, and I have to go into the app and hunt for the hidden function that realigns the hands. Which is interesting in itself, because this is clearly a known problem, and rather than fix it, they sneak a function into the app that you can use to recover from it when it happens. If you can FIND it.

Even step counting isn't terribly reliable. One study published in 2020 found that wrist-worn activity monitors such as smart watches varied in mean average percentage error from 208.6 to 861.2 per cent. That upper value means some activity monitors are recording over eight times the actual number of steps! ON AVERAGE! The most accurate was more than double! That's over a full day. Walking briskly and purposefully is a lot more accurate, so clearly smart watches are biased against the fine art of the amble. Not to mention the way they record clapping, washing dishes, and many other things as steps that are not, in fact, steps. My husband's Fitbit records him taking thousands of steps at the same time as riding his bike, which would be an impressive achievement, if only it were true. Similarly, on a recent road trip, I discovered that I took thousands of steps while sitting peacefully in the back seat of our car.

The tech industry sells us robotic vacuums that will happily share live video footage inside your home with hackers, and new cars that are so pointlessly internet connected that they can be remotely controlled using only the car's licence plate number. Tesla markets an autopilot that is in no way safe to use unsupervised. And, make no mistake, calling it 'autopilot' while the fine print says it's your responsibility to pay attention, because it's not actually an autopilot, is an incredibly arrogant way to dodge responsibility for a system that's not fit for purpose. Tech also sells us 'self-driving cars' that have two human operators per car in a control centre, monitoring them for problems. Which introduces all kinds of issues of reaction time and latency, but sure, they are self-driving. Of course they are. Tech can sell us perimeter intruder detection systems that can be fooled by boxes – giggling boxes, at that – and facial recognition systems that can't tell people of colour apart. And it can sell us blood oxygenation meters that don't work properly on dark skin.

Now it's selling us AI, with Google CEO Sundar Pichai claiming Google has a system that's showing emergent behaviour, doing things it was never programmed to do. (Spoiler alert: it does not.) Microsoft's Bill Gates says the amount of energy used by AI doesn't matter, because AI will solve climate change for us – in the complete absence of any evidence of AI ever having solved novel problems. Sure, in some very specific cases it is excellent at doing things it has been very precisely and carefully trained to do, things we already know how to do, but it has never yet solved a problem it hasn't been taught to solve by someone who has already solved it. We have no evidence that it will ever be capable of solving climate change, but every evidence that its voracious appetite for energy and water are rapidly making it worse. And there's more, and worse – AI chatbots in healthcare making up symptoms and treatments, chatbots reporting entirely fabricated criminal histories, chatbots encouraging teenagers to kill themselves. But I'm going to move on because we could list AI fails forever, and make no progress.

People really want to believe that AI is, or is about to become, sentient. Even the folks who think it's going to kill us all seem to think that would be an excellent – or at least really exciting – outcome.

Todd Sampson, on *Gruen* this year said: '[AI] is possibly the most profound transition we will definitely experience in our lifetime.'

Linda McIver, on the couch, said: 'It's really not.'

Let's have a quick poll. Are you team Linda, or team Todd?

The idea that large language models (LLMs) will magically become intelligent, life-changing, problem-solving creatures is like believing that you can drop a large amount of dry spaghetti into a bucket, and if you stir it enough, it will create a perfect scale replica of the Eiffel Tower. It's simply not how these things work. The AI industry's infinite number of monkeys will never produce *Hamlet* in real life.

But it's not the general public's fault that we are so credulous, so willing, so desperate, to believe in AI. We are literally being hyped into it. The AIs themselves are designed to maintain the illusion. I asked Claude to summarise a chunk of text into 200 words, and it produced 232 words. So I asked it why it can't count.

This is what it said. What do you notice about this response?

'I apologize for my mistakes. You're right to point out these errors, and I appreciate your patience. The truth is, I can count and I can adhere to word limits, but I made careless errors in my previous responses. These mistakes were not due to an inability to count, but rather a failure to doublecheck my work carefully. To be clear: I am capable of accurate word counting and of producing summaries to exact word limits. My failure to do so in this conversation was an oversight on my part. Thank you for holding me accountable. In the future, I will strive to be more diligent in verifying my word counts and ensuring I meet specified limits accurately.'

It's humanising, anthropomorphic language. I apologise. I appreciate your patience. I made careless errors. I failed to carefully double check MY work. I made an oversight. You are holding ME accountable. I will strive to be more diligent.

But despite the apology, let's be clear. Claude was working exactly as intended. Because LLMs are not designed to produce facts, right answers, or anything at all accurate or reliable. They are designed to produce statements that are plausible. As Lilly Ryan puts it, their statements are not facts, they are Fact Shaped.

And I would go further. Their summaries are not summaries. They are summary shaped. Their solutions are not solutions. They are solution shaped.

And the systems are not people. But they are increasingly people shaped. And this, too, is an entirely deliberate strategy. Because, as Adrian Tchaikovsky so eloquently put it in his funny and deeply disturbing book *Service Model*: 'Humans have been reading personality and self-determination into inanimate phenomena since long before Alan Turing ever proposed a test. The level of complexity in interaction required for an artificial system to convince a human that it is a person is pathetically low.'

Making them as human shaped as possible makes us more likely to want to interact with them, to believe that they are intelligent, and to pay for them to do jobs for us that they are in no way qualified to do.

The trouble is that we are not, by nature, critical thinkers. We want to believe.

Ever since the world went mad for generative AI I have been unclear as to my role. Am I Cassandra, doomed to see the future but never believed? The canary in the coal mine? The child who is the only person prepared to say out loud that the emperor is actually naked?

Whenever I sound a cautious note I get bombarded with:

'Sure, the OLD versions do that, but the latest versions are dramatically better.'

'The next version will be game changing.'

'These are the worst AIs we will ever use. The next generation is going to change the world.'

'Ah, but have you tried <insert random other LLM here>? It's different.'

And my personal favourite, this comment left on my blog:

'But, LLMs are really good at language and, once you connect an LLM to a computational system, the game changes. Just as a human has five senses, an LLM with multiple sensory inputs will quickly learn to overcome all the limitations that you parody in this post ...'

Which ... really ... I have so many questions. The idea that LLMs can become intelligent given the right inputs rests on a completely false model of how LLMs work. To summarise: LLMs ARE computational systems, and they DO NOT LEARN, even if you plug them into a nose.

I recently saw a person on LinkedIn unironically using ChatGPT to detect bogus chatbot-generated references, which is rather akin to asking that dude on Facebook whether the Rolex he wants to sell you is genuine, only worse, because the dude on Facebook knows the Rolex is a fake. ChatGPT has no clue. It is not designed to have a clue.

It has been suggested that LLMs learn from text online. That they read text and use them as inspiration. But, like Terry Pratchett's Gaspode in *Men at Arms* ('I've read books, I have. Well. Chewed books.'), LLMs don't so much read books as chew them.

It's busily being marketed as game-changing, but the little objective research that I have seen done suggests that it takes more time and effort to use than to do without it. I keep wondering if there really is some secret AI that actually works that I'm missing.

I've mostly played with text generators, so when my friend Robyn told me she'd had fun trying to get a generative art system to give her a picture of a three-legged puffin, I decided to try it for myself. It didn't matter which system I tried. It could show me a puffin, but never a three-legged one. I tried a range of prompts, starting with '3 legged puffin' and increasing in desperation through '3 legged puffin with 3 legs' to '3 legged puffin with 3 legs that actually has 3 legs' through to '3 legged puffin with 3 actual legs with the third leg on its stomach.'

But, as another stark reminder of the fact that these systems do not think, it had no way of producing a three-legged puffin. That would require some level of understanding. What is a puffin? What is a leg? It has no idea. But it has seen images labelled puffin on the internet, so it can give us something like that. It's really a pattern-matching device with a little bit of mixing. 'Here are 5 images that kind of match your prompt. The system is going to combine them into one.' It's not 'copying' copyright material. Oh no. It's using it as 'inspiration' (more humanising language).

It's not using it as inspiration, though. When you taste a cake and try to recreate the recipe on your own, that is using the cake as inspiration. When you throw the cake itself into the blender, that is not using it as inspiration, that's using it as an ingredient. Which is what generative AI does. Generative AI throws all of its source material into the blender.

A human being, having never seen a three-legged puffin, but understanding the idea of both puffins and legs, could produce a picture of a puffin that has three legs. That's using ideas as inspiration. An AI cannot, because it uses previous images as content, not inspiration. Generative AI does not, in fact, generate. It regurgitates.

AI image generation is just like the claims made about AI. At first glance it looks amazing, but when you look at it closely, the flaws are often obvious, and quite disturbing.

The entire AI industry is selling us the idea that it can do things it cannot, in fact, do. Selling us the idea that it HAS ALREADY done things (like solve problems, or teach itself new skills) that it hasn't, in fact, done. And, for most of society, the sales pitch is working alarmingly well.

Add to that cases where tech companies have acted unethically, such as the recent revelations that I-MED has sold patients' X-ray data to a startup to use for training its AIs, and it's increasingly clear that we need to rein in the tech industry's behaviour.

Technology is also wildly successful at selling us conspiracy theories, lies about health, lies about people, lies about politics, and lies about climate change, among many, many other lies.

Last year Tiktok tried to take me down a rabbit hole of videos of people trying to break into hotel rooms. I was sucked in for a while, but it really puzzled me that the women – it was always women, inevitably travelling alone – who were anxiously watching someone break into the room they were in, were filming it to put it on TikTok rather than call the police. Each video seemed to escalate in creepiness and hysteria, until I shook myself free. I now skip any video that doesn't involve cute animals (mostly cats) or David Tennant. Unless you constantly redirect it, the Tiktok algorithm seems to prioritise fear and rage.

This year I decided to try Threads for promoting my work. While exploring it and thinking about how to build a following there, I posted a small story about Doordash, and experienced my first ever nutjob pile-on. It was intense. The level of abuse and outrage was entirely disproportionate to the story. My post seemed to be getting noticed by a large number of people who weren't in my network at all. Talking to other tech friends about it, they've noticed the same things. The outraged pile-ons happen fast on Threads, even when you only have a handful of followers. They seem to have very effectively optimised the algorithm to maximise engagement using, again, fear and rage. I hardly use it now.

And as if that's not enough, the tech industry now sells us devices that will cease to function at all if the company making them goes under, or simply decides not to support them anymore. For a doorbell or a thermostat, that's annoying (and possibly expensive). For a door lock, that's a real problem. For medical technology like a bionic eye, that's a nightmare.

Technology does seem to choose to give us a lot of grief (and fear and rage). And now, not content with merely spreading misinformation, the tech industry is using AI to generate more of it. Tonnes more.

And yet. Last year technology also gave me two new hips that radically changed my quality of life. I could stand up, fully weight-bearing, as soon as I woke up from the surgery. This still astounds me, more than a year later.

Technology can be life saving. AI systems can successfully identify some types of cancerous cells faster and more accurately than radiologists.

Technology enabled the modelling of the SARS-CoV-2 virus that led to the incredibly rapid development of an entirely new class of vaccines that use novel messenger RNA-based techniques to vaccinate people against COVID-19.

My friends at Pawsey Supercomputing Centre in Perth are using High Performance Computing to develop technology that can monitor the vital signs of patients with traumatic brain injuries in real time, and predict potentially fatal intracranial hypertension before it happens, so that medical staff can prevent and treat it. Technology is saving lives.

Technology can examine aerial photographs of crops and identify the beginnings of diseases, so that farmers can treat them before they spread.

We can send scientists to space. We can photograph distant galaxies.

And, my personal favourite, my friend Darren can send me live footage of his cats in San Francisco while he's in New York and I'm in Melbourne.

Before I got up to deliver this talk, my besties in Adelaide, Brisbane, San Francisco and Perth wished me luck, and my kids several suburbs away rolled their eyes at me in real time.

So the answer is not to ditch the smart phone, or condemn technology altogether. The answer is to:

<critically evaluate technology>
<critically evaluate information>
<ask rationally difficult questions>

To do that, we need the whole of society to be strong on critical thinking, and to know enough about technology to know what questions to ask.

Schools and universities will proudly and loudly tell you that they produce graduates who are strong in critical thinking.

And the curriculum is fine as far as it goes. There are a tonne of useful concepts on there. The problem is not so much with the content as it is with the context.

As long as we're teaching using problems that have answers we can look up in the back of the textbook. As long as we're using exams to measure students' progress, or worse, to measure students! As long as we're asking all the kids to do the same thing, and assessing them on whether or not they got the right answer ... As long as we allow kids to think that their Year 12 results define them in any meaningful way ... we're not teaching critical thinking.

We're teaching compliance, and rewarding group think. We're teaching exam passing, and getting the right answer. The trouble is, in real life, with real problems, there mostly is no right answer. There are multiple solutions, each with different pros and cons. To figure out whether we've implemented a good solution, we have to critically evaluate it. Figure out who it helps, and who it harms.

When I was teaching, I first started using data science with my Year 10s because I wanted to give them a reason to learn programming. Before that we had them drawing pretty pictures with block-based languages, teaching Lego robots to push each other out of circles, and experimenting with slime mould. For some reason, they couldn't see the relevance of this to their futures. It's a mystery.

Given that my purpose was motivation, it seemed important to make it real, so that students could see how what they were doing could be useful in other contexts. We did projects on election data, on micro bats (bats, not bits), and on climate science. In every case the datasets were large, messy and complicated. Magnificently real. And in every case, the overwhelming feedback went from 'Why are you making me do this??' to 'Oooh, this is so useful! I used it in my science project, in my maths exam, and watching the news last night there was a graph that was SO misleading because there wasn't a zero on the scale ...'

It worked a treat. Kids were more engaged; they learned that coding was not only something they were capable of, it was something worth doing. And I could have left it at that. But the longer I worked with real data, the clearer it became that using real data has another benefit. There's no clear right or wrong. Let me explain.

The first dataset I taught my Year 10s with was an election dataset from the AEC. Over 3 million lines of CSV, where each line was a Victorian vote for the Senate in the federal election. Too big to open in Excel, they had to use Python to wrangle the data. We got the kids to figure out their own questions to ask of the dataset. With a dataset that big, and that complex, there were so many possible questions. It was also incredibly messy data.

We taught kids the rules of Senate voting in Australia (no small task in itself), showed them the huge ballot paper, and then opened the file to have a look. It was a spreadsheet (comma separated values) file with over 160 columns, and over 3 million rows. We'd tell the kids that people could vote above the line or below the line, but not both, and so they would write code that assumed if they found above the line voting, there would be nothing below the line. But, of course, not everyone followed the rules! The fifth line of data already has someone who has voted above and below the line on the same ballot. Surprise! People, just like data, are complicated, messy, and often don't follow the rules.

The teachers I was teaching with were kind of daunted at first, because the trouble with asking 200 students to find their own questions to ask of a large dataset is that we, as teachers, aren't going to know the answers. Which actually turns out to be one of the best parts of the whole idea – because now, instead of marking kids right or wrong, the kids have to critically evaluate their own work, and explain where it works, where it falls down, and where they aren't quite sure about it. They have to figure out what other reasons there might be for the results they found. Suddenly they have to ask themselves the question: How do I know I'm right? Imagine if we had to do that for all the projects we do, all the real problems we solve – not just in education, but in business and government!

So now we were using projects in class where finding fault with your own work is actually what gets you marks. Because we were getting the kids to solve real problems, using real data, there's no such thing as a perfect solution. There are always going to be issues. The standard educational approach is to mark right or wrong. And then we wonder why kids cheat when they don't get the results they expect in experiments and things. Because they know that the important thing is to get the right answer. What if, instead, we taught everyone that the important thing is to think critically. To evaluate your solutions. To consider other possible reasons for the results that you found.

Imagine how that might change society.

So ... we're teaching programming using data. The kids didn't all write incredible feats of machine learning as their assignments (though some did). Many of them simply wrote a page of Python to extract some data from a CSV file, and process it a little. They didn't even do the visualisations in Python, they mostly did those by hand. (Let's face it, Python visualisation libraries are not super friendly to beginners.)

But, actually, just writing a little code, and writing it successfully, was a massive achievement for some of these kids. They came into the course believing that coding was hard, not something they could do, and irrelevant to them and their futures. They dreaded it. They left the course knowing that it was not actually that hard, they were more than capable of doing it, and it was hugely relevant and useful for their futures. It turns out that learning a wide range of complex programming skills is far less important than simply learning that programming is something you can do.

Now, not only did we have a whole cohort of previously reluctant learners suddenly engaging enthusiastically with the work, we had a whole new set of kids choosing to do the Year 11 computer science subject. Including double the number of girls who had chosen it before.

I was successfully getting more kids into programming. And, let's be clear, it's not just girls who are shunning tech in alarming numbers. They're just the easily measured ones. The boys and non-binary kids who just don't think they can do it, and don't see any reason why they'd want to, are also a problem, because they, too, contribute to our lack of diversity.

But the other reason I started doing data science with the Year 10s was their science projects. Every kid at that school had to do an Extended Experimental Investigation as part of their core science unit. When I walked around at the exhibition night and looked at all the posters, the graphs frequently made me cry. They were so bad! But not only did the kids not know any better, *their science teachers hadn't spotted it either*! It was starting to dawn on me that we weren't getting basic data literacy across to the kids.

But not just the kids! You only have to live through a pandemic to figure out that very few people understand exponential growth. (So what if we had 2, then 4, then 8, then 16 cases! Those aren't scary numbers! Maybe not … but they're a terrifying pattern that will be scary REALLY SOON!)

Exponential growth is in the curriculum, and has been for a long time. So why doesn't anyone seem to understand it, or recognise it when they see it?? Because it never mattered enough to us to really learn it before.

Once you start looking for it, data ignorance is everywhere. A journalist published a graph on Twitter with no labels and no scale. When I pointed out that it wasn't actually meaningful, I got jumped on, and another journo DM'ed me to say we were all intelligent enough to know what that graph means. Except. It doesn't mean anything! It tells us there has been change, but we have no way of knowing what the change was, or what the scale might be. Without labels and a scale, that data could be going up, down, or sideways, by massive amounts or tiny, or somewhere in between.

The way he believed that 'we are all intelligent enough to know what that means' is disturbing, because too often we underestimate our ignorance when it comes to data. And we are easily swayed by some statistics, or by a pretty graph, without knowing what questions to ask to get a sense of what story that data really tells.

I have a podcast called *Make Me Data Literate* where I interview folks who do cool things with data, and I always ask them what the first question they ask is, when they look at graphs in the media.

There have been 29 episodes so far, and there are three very strong themes that have emerged in response to that question.

1. What's the source, and what story are they trying to tell?

2. Check the axes. Does the Y axis start at 0? Does it go up linearly? (And, honestly, some of the things that happen to Y axes out there in the real world should be outlawed by the Geneva Convention.)

3. What's the sample size?

Most of us don't, by default, ask those questions, simple though they are. We tend to bend at the knees when we see a graph, and assume the story it's telling us is valid. Imagine if we all knew enough to be rationally sceptical of those stories?

So ... we all need to be more data literate. More rationally sceptical. Better at critical thinking.

How do we get there?

There are a number of roadblocks along the way. When I first started doing a PhD in computer science, my husband was doing a masters in electrical engineering, specifically about windmills. When we went to parties and introduced ourselves, people would ask what we did. He'd say 'I research windmills' and people would light up. I'd say 'I'm doing a PhD in computer science' and they would physically recoil, then turn back to my husband with his nice, friendly windmills. I learnt to say, instead, that I was doing a PhD in making computers easier to use, and suddenly I was extremely popular.

Computer science is SCARY! So, too, is data science. People assume that it's hard, inaccessible, and not something that they could ever do. This is, of course, nonsense. Anyone can learn to program, and anyone can wrangle data. Just like anyone can learn to be good at maths. But we don't see it that way, because we don't teach it that way.

We have teachers who believe they are terrible with computers. Teachers who are terrified of technology, and of maths. Especially at primary school, kids believe their teachers are God.

And if God is scared of this stuff, what hope do I have?

And it's not the teachers' fault. They went through this same system. We taught them to be afraid of maths, and the tech industry taught them to be afraid of tech, when it ate their work, crashed unexpectedly, or was just bewilderingly difficult to use. When I was writing *Raising Heretics*, I did some research into primary teaching degrees. Do you know how many core units most of those degrees have on computing, or on data?

None. None at all. Clearly that needs to change. But it's no good starting at the tertiary level. We know that people's attitudes to maths, and their ideas of their own abilities, solidify in primary school. By the time they reach high school we've lost them. So we've got to go right back to the start and give kids a reason to learn this stuff, the idea that it's worth doing, and the belief that they can do it.

A lot of what I teach in my projects is already in the curriculum, but it's divorced from context. It's presented as knowledge for knowledge's sake. When I created my first data science unit, one of the teachers who had to teach it was a maths teacher, and she gave me a very hard time about teaching graphs. 'Why are we doing graphs? We already do graphs in maths. It's pointless.'

And then she saw the finished graphing unit, and she was blown away.

'This is so useful! I'm going to use this in maths! I've learnt so much!'

So. This person, with a degree in maths, who had been teaching maths for years, did not understand the real application of graphs. Take it outside of a textbook and it's useless. Sure, we can teach graphing the equation $y = x^2$. But what's the point?

We can't expect students to engage with this stuff that they 'know' deep in their hearts is not relevant to them, too hard for them, or not interesting to them. We have to give them a purpose. We have to sneak them into it sideways. Not 'We're going to teach you stuff you hate' but 'Hey, here's a real problem. How can we solve it?'

To do that, we need to build real problem-solving into the curriculum. And not just design thinking, plan-a-trip-to-Mars-type rubbish. We have to get them solving real problems, implementing their solutions, and then – and this is the really important bit – critically evaluating their solutions to see who they help, who they harm, and how they can be improved. (Imagine if governments routinely did that, hmmm?)

I have a template for this kind of problem-solving:

- Find a problem the kids care about.
- Measure it.
- Analyse the measurements.
- Communicate the results.
- Come up with a solution.
- Implement the solution.
- Measure it again to see how well it worked.

And this is the beautiful part, because the measurements will inevitably be flawed – all measurements are! (And wouldn't the world be different if we all understood that better!) What was different between the first time you measured and the second, aside from what you did? Was the weather different? Were people away? Was the traffic lower due to a rostered day off on the local construction sites? What other reasons could there be for the results you found? How could you test for that?

And then you write a report about what you found. Or you write a letter to the school leadership advocating for change, with evidence! Or you write a submission to the council, or the local public transport provider, or the state government.

You can cover nearly every subject in the curriculum with a project like that. English for the communication, maths for the analysis and graphing, art for the visualisation, science for designing experiments to measure the problem, and potentially for topics around the problem itself, geography for the people and place aspects of the problem, history for the origins of the problem. And computing all the way through.

And that's why I created the Australian Data Science Education Institute, or ADSEI. ADSEI is a charity dedicated to building the critical thinking, data literacy, and STEM skills of all Australian students, from childcare to university. We design projects based on that template, and also projects to analyse existing datasets on any topic you can think of. We curriculum-link them, and train teachers in the data skills necessary to teach them. A lot of that is confidence building, really. We have learned to be afraid of maths, afraid of data, and particularly afraid of technology.

Some example projects include exploring the difference between a single house installing solar panels, and an entire community doing so (individual versus collective action). An exploration of pocket inequality, where kids count the pockets in men's, women's, and unisex clothing, and evaluate their effectiveness (thus accounting for those uselessly shallow pockets that actually entrap you into dropping your phone). A study of the differences in sleep habits of different age groups. Collecting data about how much further people have to go to use the 'accessible' features of a venue, and then reporting on it. Evaluating the impact of mindfulness and wellness apps. Evaluating the accuracy of different weather prediction apps. Collecting data about the usage of different social/ activity areas in the school grounds and coming up with strategies for democratising access to them.

One of my former students, who now works with data professionally, once told me it was because of me that she was in this job. 'You're the one who gave me the confidence to play with data like silly putty.'

Giving teachers, and hence students, that confidence is a huge part of what I do.

But it's also very much about the critical thinking.

If kids don't have right or wrong answers, and can't look them up in the back of the textbook, then they have to learn to critically analyse their own work. Suddenly we're not rewarding right answers, we're rewarding them for identifying issues with their work. So you take out the defensiveness, the obsession with perfection, and the need to be right, and you replace it with the knowledge that your work isn't perfect, and that identifying how it's not perfect is the thing that matters.

Just pause for a moment and consider the difference it might make to the world if we didn't learn we had to be perfect. If we learned we had to be thoughtful instead. If we learned that nothing is perfect, even our own work, and everything needs to be critically evaluated.

Neal Stephenson, in *Diamond Age*, wrote: 'The difference between stupid and intelligent people – and this is true whether or not they are well-educated – is that intelligent people can handle subtlety. They are not baffled by ambiguous or even contradictory situations – in fact, they expect them and are apt to become suspicious when things seem overly straightforward.'

I quite like this definition, and it means that by teaching us simple, neat, perfect scenarios, our education system is actively making us stupid.

Once I realised that these projects could actually teach kids to critically evaluate their own work, I stopped caring about teaching them programming. I don't care whether they use Python, Tableau, Excel, pen and paper, or blocks. It's the critical thinking that matters. It's the values that we're teaching them.

Because assessment inevitably teaches kids what we care about, and for too long it has taught them that we care about right answers. About memorisation. About regurgitating facts and applying known processes. Rather like AI, now that I think about it … When we could be using assessment to teach them that we care about creative problem-solving, critical thinking, and identifying the problems with their own work. We could be teaching them that they have the power to change the world. That what they do matters.

Instead of using assessment to teach kids to follow rules and produce the right answer, let's use it to teach them to be rational sceptics, critical thinkers, and absolute troublemakers. Because troublemakers are the ones who are creative enough, and radical enough, to solve the hard problems, and to change the world.

So, if I figured out how to engage kids with STEM, how to teach critical thinking, how to motivate kids to learn programming, why am I not working with kids? Why work with teachers instead?

Well, in a nutshell, my goal is to put ADSEI out of business. I recognise that this is a weird thing for a founder to say. But if I run after-school programs, or school incursions, or anything else that has me working with kids, it doesn't scale. If I work with 30 kids, I've worked with 30 kids. But if I run just one workshop with 30 teachers, each of whom has up to 12 classes per year – 6 per semester – of 25 kids each, I've worked with up to 9000 kids. PER YEAR.

And it's more than that, because my goal is systemic change. Change the way teachers teach, change the curriculum, change the teacher training. Change the whole system. So ADSEI also works in the policy space, advocating for change, advising on curriculum, and generally being the kind of troublemakers that we say the world needs.

When you design a new course, best practice is to spell out very clearly and explicitly what the purpose of the course is. What we really need to do now is to redesign the education system, and to spell out very clearly and explicitly what the purpose of education is. What are we trying to do here?

Here's what I want our education system to do. I want it to produce:

- critical thinkers
- creative problem-solvers WHO CRITICALLY EVALUATE THEIR OWN SOLUTIONS

- challengers of the status quo – people who ask WHY
- evidence-based policy-makers.

People who look at 'smart watches' that can't tell the time, and chatbots that can't tell the truth and ask 'Why?' People who say 'Do better', and know how to measure what 'better' looks like.

I've put this in the context of the tech industry, but we can apply it to medicine, to government, and to business, just as easily. We absolutely need to apply it to climate change, pollution and species extinction. We need to apply it to the housing crisis, to income inequality, to politics, and to media.

I put my blood test results from the last 12 months into a spreadsheet recently and figured out patterns and issues that were invisible to my GP, who was comparing one test to another by scanning the awful printouts that pathology labs send. Why aren't we teaching our doctors to manage data systematically? How many developing issues go unnoticed, because all they know how to do is scan the printout for items in bold that are outside the 'expected range'?

What we can build this way is a world where policy is evidence-based. Where we make data-informed decisions, while understanding that the data isn't perfect. Where kids are empowered to learn all of the skills they need to solve problems in their own communities. Where technological solutions are rationally evaluated, rather than uncritically worshipped.

Imagine the world we could create.

✳ *Gaza: Why is it so hard to establish the death toll?*, p. **169**
Stanford prison experiment: Why the lead psychologist defended his infamous study to the end, p. **282**

SKINK ON THE BRINK

Anthea Batsakis

The wintry forests of the New England Tablelands sit high in New South Wales' Great Dividing Range. Magnificently lush and studded with waterfalls, these are the final bastion for many of Australia's rarest species.

The last wild regent honeyeaters flit between ironbark gums. Scruffy Hastings River mice shelter in hollowed logs. And hidden in the southern edge of the Walcha Plateau – in the traditional land of the Biripi People – lies an animal that may well be the next in Australia to go extinct: the long sunskink.

The long sunskink is, frankly, an unremarkable little reptile. At only 5 centimetres long, it can easily fit in the palm of your hand. It's brown and bronze, with a cream-coloured belly and a snake-like body. To the untrained eye, it looks almost exactly like the delicate skink, one of Australia's most common reptiles. It is this inability to capture the imagination that may be the skink's downfall.

The long sunskink was first collected in 1972. It inspired such little interest in the academic community that no one bothered to scientifically describe it for another 25 years. And for nearly nine years – between 2014 and 2022 – no one could find it.

But this wasn't acceptable for Jules Farquhar, a research associate at Monash University. He and his honours student Kelsey Graham conducted a gruelling, 'blitzkrieg' field survey in its last known habitat.

No stone unturned, literally

It's not easy to find an animal when almost nothing is known about it. The long sunskink (*Lampropholis elongata*) is listed as Data Deficient (DD) on the IUCN Red List – the international database for the extinction risk of the planet's threatened species. As Jules puts it, data-deficient species 'are often just threatened species in disguise, waiting to be assigned conservation status'.

In fact, 14 per cent of the world's skink species are considered data deficient, and another 8 per cent remain entirely unassessed. But at Monash University, a research group is dedicated to changing this in Australia. Led by ecologist David Chapple, the researchers keep a spreadsheet of data-deficient skinks in Australia. One by one, and with enormous effort, each skink is rescued from the deadly realm of obscurity.

The long sunskink was one such species. In the spring of 2022, Jules and Kelsey spent five weeks scouring Ngulin Nature Reserve and the Riamukka State Forest in the New England Tablelands. Battling dense vegetation, they checked beneath every log and stone, set up traps, counted rocks, raked through leaf litter and prised open tussock grass.

'You're looking for a needle in a haystack as it is, and when you do see the flash of a lizard, it's probably one of the tens of delicate skinks we saw per day. But we still had to dive into the bushes to check,' Jules says.

It was painstaking work. Jules has a sunny temperament, a quick laugh and, also being a snake catcher, is no stranger to wrestling reptiles. Yet he says he spent months after this intense fieldwork recovering. 'It was particularly challenging because the soil there is incredibly hard and rocky, with thick vegetation. I absolutely burned myself out on that trip,' he says.

Jules and Kelsey left the forest each day empty-handed, with the frightening probability of extinction creeping higher. Eventually, however, they discovered that the last places the skink was seen in the years prior were no longer the right places to survey.

Eureka!

With only a week of their survey remaining, Jules and Kelsey stopped by a small, sunny clearing to photograph water skinks known to that area. Jules casually turned over yet another rock and there, just 23 millimetres long, was a baby long sunskink.

'We were unsure, because it was such a small individual. So I flipped another rock five minutes later and found a full adult. I made sure I was certain of the species, because I knew what I was about to yell out. I held up a hand in the air and shouted, "Hey, Kelsey, I got one".'

Jules says they got the data they needed and released it back into the wild. Then, they 'made a beeline for the pub'.

Long sunskink habitat

After that, compared to the previous weeks' slog, finding long sunskinks became a breeze. This time, the researchers knew to target clearings in the forest with connected clumps of tussock grass, where the little skinks can travel in the shelter of its blades on rainy days.

'That situation is actually quite rare in the forest,' Jules says. 'The skinks are really fussy with their habitat. We may never know their story and how they got to this position. But I just assume they evolved themselves into a corner.'

In total, Jules and Kelsey found 23 long sunskinks in four different locations.

A skink on the brink

Prior to the duo's research, the long sunskink wasn't listed under Australia's wildlife conservation law.

Now it is officially listed as Critically Endangered, joining more than 2200 plants, animals and ecological communities on Australia's threatened species list.

This is a major achievement, because it means there are now legal requirements to protect it. But it is still only the first step in preventing its demise, and doesn't guarantee its safety.

From feral pigs to climate change to logging, threats to the long sunskink abound.

Indeed, all four different clearings found to contain the skink are within land available for harvesting by the Forestry Corporation of NSW. They're also right beside a Forestry Corp pine plantation. The skinks are so close to the road dividing the forest and the plantation that Jules saw four-wheel drives miss the skinks by only a couple of metres. With so few long sunskinks left, every death is a major loss to the species.

'They're on the razor's edge of this road. I also worry about logging machinery running over the skinks, not just removing its habitat,' Jules says.

Jules alerted Forestry Corp to the skink's proximity to its operations in 2022. *Australian Geographic* contacted the company to learn what measures it has put in place to protect the skink's survival since then. A Forestry Corp spokesperson said over half of Riamukka State Forest is permanently protected for conservation, and that timber harvesting takes place only in areas that have been harvested and regrown previously.

The spokesperson said that in timber plantations, the independent regulator must assess its 'unique and special wildlife values' and the measures put in place to protect them. In native forests: 'If the species is found in an operational area or within close proximity to the boundary, additional site-specific biodiversity conditions must be developed and approved by the independent regulator.'

Jules says he is concerned Forestry Corp's protection measures are only contingent on whether a long sunskink is detected – and they're not exactly easy to find. 'My hope is that Forestry Corp is actually doing pre-harvest surveys that are capable of detecting the long sunskink if it is present at a site,' he says.

Habitat disturbance from logging is also deepening the wounds that other threats are inflicting on the species. For example, feral pigs and deer tear up the soil in the forest, particularly along the drainage lines that intersect the clearings. Given the skinks heavily depend on native plants, invasive weeds could also see them unable to cope. And bushfires, worsening under climate change, could destroy a local population in one fell swoop.

'Fire is an extreme threat. I have a map of the extent of the 2019–2020 bushfires. I plotted it against the cluster of the species, and I nearly spat my tea out,' Jules says. 'It was all red, all around the species – a single fire event could have easily wiped out the entire distribution.'

A seat at the table

The long sunskink is just one example of the hundreds of 'boring' species on the brink of extinction – many of which aren't protected under federal law because we don't have the data to prove their rarity.

Skinks, and many other reptiles, have the added problem that most live in remote deserts, out of sight and out of mind.

David Chapple leads the Monash University research group on data-deficient skinks. He says the standard situation for a native skink species is to not have any dedicated study on it since it was first described. 'It means we're lacking a lot of fundamental information, like what its habitat is like and whether its population is declining,' David says.

Australia doesn't even have a centralised, national list of data-deficient species, unlike the IUCN Red List. David says this is a problem because at the federal and state level, 'any animal not listed as "threatened" is considered to be of least concern'.

However, research this year shows that Australia's data-deficient snakes and lizards are three times more likely to be on the brink of extinction than those assessed. A national list could be the first step in giving them formal protections, and raising awareness for their research needs. 'Data-deficient species don't even have a seat at the table to be considered for protection. Even though we know they have a high likelihood of being threatened, they're lost and overlooked in the entire process when there's no requirement to protect them.'

The long sunskink, then, may actually be one of the lucky ones: we now know it exists, what threatens it, and how to protect it. Now it needs money and boots on the ground.

'In order for those things to happen, there needs to be awareness. People aren't going to put time and effort into something they're not aware of,' David says.

✳ *Cat-astrophe: Australia's feral cat problem*, p. **33**
'Earth poetry' in the Arctic, p. **53**

MY PARTNER AND I BOTH HAVE LONG COVID. WE TREAD THE UNDERWORLD TOGETHER

Felicity Nelson

There's a story in Greek mythology where Orpheus, a gifted poet and musician, travels to the underworld to rescue his newly wedded wife, Eurydice, who died after being bitten by a snake.

They are both allowed to leave Hades under one condition: Eurydice must walk behind Orpheus, and he must not turn and look back at her on the way out.

I told my partner this story late one night. He hadn't heard it before, but he saw the parallels with our story right away.

I caught SARS-CoV-2 at a work Christmas party in December 2021, a few weeks after my partner and I had moved in together. We were travelling from Sydney to Melbourne when we heard that someone at the party had tested positive.

We were among the first people in Australia to get this disease that had caused millions of deaths worldwide. I was terrified.

At that time, Australia's state governments had very strict quarantine rules, so we couldn't cross the border and go home to Sydney.

My partner wanted to care for me while I was ill, so he stayed in the same house as me instead of quarantining separately. A few days later, he fell ill too.

I had classic cold and flu symptoms, was breathless, and lost my sense of taste and smell.

My partner's COVID-19 symptoms were much more severe

than mine. He had a fever and extreme muscle aches. We called the ambulance out when he was struggling to breathe, but the paramedics decided that he did not need to be taken to the hospital because his blood pressure, heart rate and oxygen levels were OK.

We had to stay in Melbourne for 16 days until my partner had recovered enough to fly home. It was such an ordeal that my diary of the experience was published by the national media.

Two and a half years on, my partner has severe long COVID, with fatigue, post-exertional malaise, brain fog, dysautonomia, sensory hypersensitivity, aphasia, paralysis, and low social tolerance.

He used to be extremely fit. He would cycle to work and back and go for long runs in the rain. But he hasn't been able to do that since getting COVID-19.

For an athletic corporate lawyer who relishes hard intellectual and physical work, long COVID is a special kind of torture.

In some ways, it's fortunate that I'm down here in the underworld with him. I can feel my way through the darkness with first-hand experience of the terrors he is facing.

After getting COVID-19, my symptoms of fatigue, breathlessness, brain fog, congestion and post-exertional malaise never went away. I still can't do the cycling, jogging or social activities that I used to, and I can only work part-time.

The thing that sends my partner back to long COVID crash hell is social activity – particularly with the people he loves. Sometimes it only takes a few minutes of the wrong kind of social strain and it's like he's been mauled by a three-headed dog. He goes from holding a conversation perfectly well to turning grey, losing his balance, wobbling and crashing into walls. If he doesn't lie down within seconds, he will slide down the wall and collapse on the floor.

My partner relapsed at the start of this year. He lost all the gains he had made in the past two years, going from 40 hours of work each week to zero. His 'rock bottom' involved the most intense sensory overload he's had since getting COVID-19.

His migraine was so bad, he was howling in pain. He was nauseated. His skin was on fire. He told me he had lost that feeling of hope that he'd been clinging onto since getting COVID-19. Hearing this from the most cheerful person I've ever met was devastating.

I realised we needed to change the way we were operating. I wrote a list of all the things he was doing, and identified the activities that were using up energy without being restorative.

Then, I identified the tasks that I could do for him – cooking, cleaning, laundry, shopping and facilitating medical appointments – and from then on, I did all these tasks or delegated them to others.

For this to be sustainable and not breed resentment, I needed breaks – or at least the illusion of taking breaks.

It slowly dawned on me that we needed to educate our families on how they could make their love accessible for someone who, even on his best days, could not socialise for more than a few minutes.

When you have lost the ability to generate hope for yourself, family support provides that hope for you. Even if you don't recover, they still love you and they are doing OK. And that's hopeful.

I asked both our families to go out of their way to share happy, normal family messages that did not require a response from my partner. The family chats filled with pictures of dogs being dopey, smiley holiday photos, and the latest sport gossip. Our freezer was suddenly chock-full of homemade meals. Our breadbox was crammed with baked goods.

The house filled with helpful sprites volunteering to cook, do the dishes, water plants, vacuum, scrub the porch tiles, mow the lawn, dust, and fix things.

It was embarrassing at first, but weeks passed, and we fell into a routine.

In the myth, just before he exits the underworld, Orpheus looks back, and his wife plummets down into Hades again.

That's a mistake I refuse to make. It's natural to want to

connect with the people we love, but that impulse can be harmful during a long COVID crash.

It's now been 12 weeks since we made all these changes at home to support my partner. He's no longer bed-bound. He can go for a few short walks, sit in the sun, and work from home for a few hours each week.

We're going to maintain support for as long as he needs. He will slip into the underworld again. Crashes are inevitable. But now we know the way out of its cold, dark halls.

✱ *How surviving a deadly animal encounter is not the end of the trauma*, p. **112**
Moments of kindness in a regional hospital, p. **258**

MOMENTS OF KINDNESS IN A REGIONAL HOSPITAL

Michael Leach

i.
a lift closes—
the orderly's hand
      sticks out

ii.
Covid restrictions
the ward nurse still lets us see her

iii.
I take a seat
to hear bad news—
the intensivist
squats

iv.
her deathbed
the music therapist plays
'Landslide'

v.
five minutes after
      she passed—
the chaplain's soft voice

MYSTERY SEISMIC WAVES THAT RIPPLED AROUND EARTH FOR NINE DAYS CAUSED BY GREENLAND TSUNAMI: STUDY

Carl Smith and Peter de Kruijff

One morning in September 2023, Rudolf Widmer-Schnidrig was doing a spot check of earthquake recording equipment when he saw something extraordinary.

The devices, at Germany's Black Forest Observatory down a repurposed silver mine, detect the most minute rumbles of the planet, allowing seismologists like himself to see signals from earthquakes or other geological events.

But on that day, Dr Widmer-Schnidrig said, 'there was a signal which looked different from anything I have seen in 22 years working here at the observatory'. There was a large burst of activity, erratic squiggles on a graph. And bizarrely, the signal repeated every few hours for nine days. As it repeated, the signal also tapered off extremely slowly, far slower than expected from something like an earthquake.

The signal could only be caused by something big, but there are very few things we know of that could've produced it.

It might have been a sign of volcanic or tectonic instability or an unknown weapons test. The other exciting possibility was it might have been some rare geological process science had not yet described.

The signal was detected around the world – as far away as Australia and Antarctica – seemingly causing the entire planet to 'ring' like a bell.

Dr Widmer-Schnidrig and dozens of seismologists jumped onto online forums to share data and speculate on what happened. 'This is really a very extraordinary occasion to work on such a signal,' Dr Widmer-Schnidrig said.

A year on, the scientists unravelled the mystery and reported their findings in the journal *Science*.

International collaboration

Scientists from 15 countries and 40 institutions pooled their resources and data to figure out what caused the mysterious signal.

Seismologist and study co-author Stephen Hicks of the University College London said with little idea of the cause initially, the researchers referred to the signal as a USO, or 'unidentified seismic object'.

'It was really like putting a lot of puzzle pieces together,' Dr Hicks said. 'I feel like it's similar to those air crash investigation documentaries. One of the first results we had showed the signal was coming from East Greenland.'

At the same time the seismologists discovered the signal, the Danish military received reports of a tsunami in the narrow, remote Dickson Fjord, which wiped out a dog sled patrol and research base that was unstaffed at the time.

So they sent a ship to collect evidence of what had happened.

The military found damage to a glacier that could only be caused by a 200-metre-high mega-tsunami in the isolated fjord – one of the highest tsunamis in recent history – 72 kilometres from the island base.

'[The] tsunami was a smoking gun,' Dr Hicks said. 'We realised, working with our Danish colleagues, that there's been a huge landslide, so a mountain [called Hvide Støvhorn] had collapsed, rolling down into a glacier. That landslide was just able to pick up more momentum, more material from the ice, and then make a giant splash into the water.'

Dr Hicks said this caused quite a bit of damage in the fjord, including to some archaeological heritage sites in the area.

'Tourist cruise ships often travel up into these particular fields and thankfully at the time there were no cruise ships in the area, otherwise the results could have been far more devastating,' he said. 'So then the mystery was: "OK, we know a tsunami has happened. But why is this seismic signal still persisting for nine days?"'

Piecing together satellite and drone imagery, before-and-after photos and data, and simulations of the tsunami, the team figured out the signal was caused by a phenomenon called a seiche. Seiches happen when water is pushed from one side of a body of water, such as a lake, to the other, and rebounds back and forth. A smaller-scale example is how water sloshes back and forth when you get into a bathtub.

In this instance, the bath was Dickson Fjord, which is 2.7 kilometres wide and closed at one end, and the rockslide, the bather.

Through complex mathematical simulations and models, the study researchers calculated there was the initial 200-metre-high wave, followed by subsequent waves of about 110 metres, which then stabilised into a seiche with 7-metre waves.

It was that seiche in the fjord that made seismometers around the world, such as those in Dr Widmer-Schnidrig's subterranean observatory, go haywire.

After further analysis, the team figured out some 25 million cubic metres of rock and ice had fallen into the fjord to create the first powerful wave.

Hrvoje Tkalčić, a geophysicist at the Australian National University who was not part of the study, said it was fascinating to see how seismology and other geophysical sciences were used in the research.

'Not long ago, global seismology was only associated with studying earthquakes,' Professor Tkalčić said. 'These days, we see a transition ... to a modern scientific discipline that uses the accumulated knowledge to study the whole Earth as a planet.'

But even after the study researchers solved the seiche mystery, questions still remained: what caused the landslide? And would it happen again?

'Ringing' Earth is the sound of climate change

After months of modelling and simulations, plus developing a precise mathematical formula to show how the water sloshed back and forth in Dickson Fjord, the research team turned its attention to what caused the landslide.

Dr Hicks said climate change contributed to the mountain collapse. 'This glacier at the base had been thinning by up to 30 metres over the past couple of decades,' he said. 'And so that mountain just wasn't able to be supported anymore.'

The study authors pointed out that accelerating climate change would likely cause more massive slides and tsunamis, potentially putting people at risk. 'You could theoretically have similar scale events, particularly as climate change continues,' Dr Hicks said. '[There are] tourist cruise ships going up into these fjords [and] other polar fjords worldwide which are more populated. You might only have to have a small instability to really kick off this larger chain of events that evolves into something that can be catastrophic.'

Professor Tkalčić thought the world would see similar seiche events in the future for two reasons. 'Glacial melting will increase the rate of ice-related landslides, and due to a unique icy landscape and topography in this part of the world [Greenland], this can lead to tsunamis and seiche phenomena,' he said. '[And] the number of seismic sensors is increasing, which results in more detections of remote natural phenomena we were not able to detect several decades ago.

'Get ready to be surprised by the plethora of behaviour[s] the Earth has in its store.'

Dr Hicks also said the September 2023 seiche showed it was important to consider climate impacts beyond the atmosphere or our oceans. 'For the first time this event is showing us that to look at the impacts of climate change, we should also be looking beneath our feet,' he said.

A watchful eye

Back at the Black Forest Observatory, Dr Widmer-Schnidrig continues to do his equipment checks, ready to capture any new mystery signals from a changing world.

'If you want to study the Earth's interior, if you want to understand the planet on which we live, you cannot start recording when the [event] has happened,' he said. 'When something happens, like this event in Greenland, then you're ready and you have it recorded.

'We should really try the best we can to understand the planet we live on.'

✱   *Mathematical transformations: The power of vectors and tensors*, p. **20**
Fishing for a glacier's secrets, p. **58**
Sounds of the slow-rolling sea, p. **175**

THE WORLD HAS BEEN ITS HOTTEST ON RECORD FOR TEN MONTHS STRAIGHT. SCIENTISTS CAN'T FULLY EXPLAIN WHY

Tyne Logan

One of the world's leading climate scientists says the world could be in 'uncharted territory', with the researchers unable to fully explain why the world has been breaking heat records to such extremes for ten months straight.

Last month (March 2024) was the hottest March on record, marking the tenth month in a row to reach that title, according to the European Union's key climate service Copernicus.

In Europe, the temperature for March was 2.12 degrees Celsius above the historical average, marking the second-warmest March on record for the continent. Around the rest of the globe, temperatures were furthest above average over parts of Antarctica, Greenland, eastern North America, eastern Russia, Central America, parts of South America, and southern Australia.

The continuation of record-breaking heat comes after 2023 was officially declared the hottest year on record, by a long way.

NASA's senior climate advisor Gavin Schmidt says while climate change and the onset of El Niño explain a significant portion of last year's heat, together with other contributing factors, there is still a margin of heat at the top that can't be explained.

He said that was concerning.

'If we can't explain what's going on, then that has real consequences for what we can say is going to happen in the future,' Dr Schmidt said.

Predictions 'failed ugly'

For about a decade, he and other climate science institutes have been making predictions of global temperatures for the year ahead. Each has a slightly different method for doing this.

Generally, it's done by looking at the baseline of global warming that the world is starting the year on, and then factoring in known climate drivers. But all of those predictions for 2023 fell short of what occurred – the closest prediction was still almost 0.2 degrees Celsius off the mark.

It may not sound like much, but Dr Schmidt said in the context of the world's climate, it's huge. 'Those predictions, based on what was happening right at the beginning of the year, failed ugly.'

Dr Schmidt said there was always room for error, but usually scientists could explain what occurred upon looking back at the data. He said this time it was not adding up. And the climate models were giving them no answers either.

'It means there's something missing in what we're thinking about here,' he said. 'Either something has changed in the system and things are responding differently to how they responded in the past, or there are other elements that are happening that we didn't take into account.'

What are the possible explanations?

Scientists have been investigating several different possible explanations for the higher-than-expected global heat.

Air pollution

Among them is the theory that the amount of air pollution around the world is less than what the climate models had been accounting for, thanks to new international shipping regulations. Many aerosols act like a 'shade' to incoming sunlight, reflecting it into space. So, fewer of them would have a warming effect.

But Dr Schmidt said, while it made some difference, it didn't seem to be enough to explain just how hot it had been. 'When you put that into a model and you say, "Is that warming effect large enough to give you this the big difference between 2022 and 2023?", the answer is no, not really as far as we can tell,' he said.

The underwater volcano

Another factor that has been looked at is the underwater volcano Hunga Tonga-Hunga Ha'apai eruption in January 2022, which shot ash and other particles more than halfway into space. Similar to pollution, volcanoes generally have a cooling effect.

But the Hunga Tonga volcano was different.

Because it was an underwater volcano, it also ejected a significant amount of water vapour – a strong greenhouse gas – into the stratosphere, and therefore is thought to have had a net warming impact.

Dr Schimdt said from what they could tell so far, this still only represented a very small change, overall. 'The magnitude of the change is in the hundredths of a degree level, so not commensurate with the size of the thing we're trying to explain,' he said.

The solar cycle and other explanations

Some have looked to the solar cycle for an explanation, as it is reaching solar maximum – something that can also have an impact on surface temperature. Solar maximum refers to the period of greatest solar activity during the sun's 11-year solar cycle.

But again, Dr Schmidt said it was not large enough to explain what they had seen in 2023, and it was 'baked into the calculations' anyway.

'And maybe it was just random things happening in the Antarctic, and in the North Atlantic, all at once, that were unconnected and are adding up, and the reason we haven't seen it before is because we haven't had 200 years of good data,' he said. 'We're looking into those kinds of things as well.'

A previous climate mystery

A similar climate model mystery has played out before, according to Dr Schmidt.

In the early 2000s, the trend of rising surface temperatures appeared to plateau for over a decade, despite greenhouse gas emissions in the atmosphere reaching record levels. It was something that climate scientists couldn't fully explain at the time, becoming known as the 'global warming hiatus'. It was also used heavily by climate change sceptics as evidence that the earth wasn't getting much hotter.

However, later studies revealed there was no hiatus in global warming – rather, it was being buried in the deep layers of the oceans. Minor revisions to data inputs, including the uptake of heat by the oceans, natural variability and observations helped make that clear.

Dr Schmidt said it was possible something of a similar nature was happening this time too, and that the climate models were missing something, or the data wasn't quite right. 'Perhaps we haven't fully characterised the Hunga Tonga volcano, or perhaps we haven't been tracking appropriately the emissions from China, because they're not necessarily the most trustworthy of global reporters,' he said.

He said it's important they work it out so they could tell whether this was simply a blip, or the start of something different.

On this, he said the global temperatures during the northern hemisphere summer could give them some clues.

All eyes on northern summer

It wasn't just the hottest year on record: ocean heat also reached its highest level.

So far, the heat of 2024 has been largely in line with expectations, according to Dr Schmidt, because scientists expect a boost to global temperatures a few months after the peak of El Niño.

But he said if everything was behaving as normal, it would cool down by June.

'The key will be what happens in the next few months. If things stay super anomalous then we are looking at a systematic change, not just a blip,' he said.

In the meantime, he said they will be re-examining data sets, including looking at newly available aerosol data from a recently launched NASA satellite, to try to explain the gap.

✱   *Air conditioning quietly changed Australian life in just a few decades*, p. **98**
    A snowflake in hell, p. **131**

WHY SOME VENOMOUS SNAKES CAN BITE AND KILL EVEN WHEN THEY'RE DEAD AND DECAPITATED

Belinda Smith

While recently investigating a branch of his family tree, Ray Miller found a newspaper story about his great-great-aunt Annie Whitby, who lived near Gunnedah in New South Wales.

The article, in the 23 January 1953 issue of the *Daily Telegraph*, was titled 'Severed head of snake kills woman'.

'Straight away I was like, "Oh, OK. That's already interesting",' Mr Miller, of Townsville, says.

In short, Annie's husband shot a large brown snake in their backyard with a shotgun, blasting its head off.

Their dog then picked up the decapitated head in its mouth. When Annie pried the head out from between the dog's teeth, 'the head "bit" her', according to the report.

Despite rushing to Gunnedah Hospital for treatment, Annie died around nine hours later. The dog, which was also bitten by the snake, died too.

This isn't the only instance of a dead snake biting a person. You don't have to look too far to find reports of people in Australia and abroad copping a dose of venom from decapitated snakes.

So how can this happen, and how long can snakes remain dangerous after they've wriggled off this mortal coil?

Why a dead snake can bite

To understand how a dead snake can be deadly, it's worth knowing why they can be such effective killing machines when they're alive.

The composition of snake venom itself varies between species but you can think of it as a complex cocktail made of molecules – up to 30 different proteins and peptides – that each might inflict different types of damage, University of Queensland snake expert Christina Zdenek says.

They're broadly categorised into three types:

- **Neurotoxic molecules.** These disrupt nerve function. Effects include drooping eyelids and being unable to breathe, because your diaphragm muscles can't contract.
- **Haemotoxic molecules.** They disrupt our blood-clotting ability, potentially leading to internal bleeding, organ failure and bleeding in the brain.
- **Cytotoxic molecules.** These encourage our body cells to burst open, releasing their contents into the blood. This can overwhelm the kidneys, which filter the blood, causing kidney failure and death.

Other toxins, such as those found in some spitting cobras, inflict pain.

This lethal cocktail is stored in a gland in the snake's cheek, which is connected to their fangs via a duct. When a snake bites, perhaps in self-defence or to subdue prey, compression muscles squish the gland. Venom courses through to the syringe-like fang, and into (or onto) their target.

Given these 'envenomations' require a muscle to contract, how do dead, decapitated snakes manage it?

It boils down to the fact that snakes are energy-efficient creatures, Dr Zdenek says. 'A snake at rest might only breathe five times a minute, so already they have low oxygen requirements.'

This means even when their heart has stopped beating, their tissues can retain enough oxygen to allow nerves to fire, triggering a bite reflex if you put a finger in or on its mouth.

'It's unsurprising to me that they can still have these post-mortem reflexes for quite a long period,' Dr Zdenek says.

For how long are they dangerous?

We humans have known that snakes appear to 'live on' after death for centuries. Back in the 1300s, Egyptian scholar Al-Damiri wrote that 'a dead viper's body will continue to wriggle for three days'.

More recently, there's the commonly held belief that 'a snake won't die until the Sun sets'.

While both exaggerate matters slightly, we have a reasonable idea of the dangers of a headless venomous snake from a series of rather gruesome experiments conducted in the early to mid-1900s by US herpetologist Laurence Klauber.

He chopped the head off live snakes (this happened before animal ethics was a serious consideration) and, among other things, measured how long after death the snake could bite a stick placed in its mouth.

'One head bit a stick and discharged venom 43 minutes after decapitation,' Dr Zdenek says.

That chomp, Mr Klauber wrote in his 1956 book *Rattlesnakes: Their Habits, Life Histories, and Influence on Mankind*, 'was certainly powerful enough to have imbedded [sic] the fangs in the hand and could well have caused a serious accident'.

As some snake toxins are incredibly stable molecules, and don't easily deteriorate, they can cause problems not minutes but months down the track. For instance, in the 1980s, a 22-year-old man went to emergency after a fang from a dead rattlesnake punctured his finger. But the snake wasn't freshly dead – far from it. It had been dead for seven weeks, during which time it was frozen, then preserved in salt, and then dipped in glycerin.

The man was in the process of mounting the snake head when one of its fangs dislodged from the animal and lodged itself in his finger.

According to the case report, published in the *Annals of Emergency Medicine* in 1986, 'the fang remained in the finger for possibly three minutes before being noticed and removed by the patient', during which time some residual venom made its way into his body.

He was treated and survived, but he felt dizzy and weak, his hand and forearm swelled up, and his hand and face felt numb and tingly.

Advice for handling dead snakes

Put simply: don't, if you can help it.

Or – and this is advice from snake-chopper Mr Klauber – at least wait until rigor mortis sets in, by which point compression muscles will lose their ability to squeeze venom from the glands.

Whatever you do, stay away from the fangs – and don't kill snakes, Dr Zdenek says. It's illegal, and snakes are ecologically important to keep around. 'And it also just puts you in a dangerous position in the event that you are bitten by an alive or dead snake.'

In a letter to the *New England Journal of Medicine* in 1999, two doctors in Arizona wrote about five cases of rattlesnake retribution, where the snakes, which had been bludgeoned or shot, bit the blokes who attacked them.

If you are unlucky enough to be bitten by a dead snake, first aid would be the same as if it was alive, Dr Zdenek says. 'You would stay calm, stay still, and wrap that whole limb with a compression stretchy bandage. That's called the pressure immobilisation technique.

'Then get in touch with medical professionals from there.'

A MUSEUM HEIST 70 YEARS AGO IS STILL CAUSING A FLUTTER IN BUTTERFLY SCIENCE TODAY

Olivia Congdon

Lepidopterist Michael Braby is sitting in his lab, peering at a photo of a butterfly specimen on his computer screen – just as he's done countless times before. He has no inkling that he's about to discover Australia's greatest taxonomic fraud. But he has noticed something odd about this particular butterfly, a Flame Hairstick (*Pseudalmenus barringtonensis*).

He zooms in to take a closer look. And on the distinctive 'flame' of the butterfly, the scarlet patch on the dark brown hind wing, there it is: it looks like it's been patched up with red paint.

Surely, he thinks, it can't have been tampered with?

Associate Professor Braby, who is from the ANU Research School of Biology and the CSIRO, has spent 30-odd years researching butterflies and moths, so when alarm bells start ringing about this suspect-looking specimen, he follows up on it, calling his colleague.

'I thought it might have been an accident,' Braby says, reasoning that someone may have guiltily tried to fix a damaged wing. But straight off the bat, his colleague Rod Eastwood makes a suggestion: it's got to be connected with the Colin Wyatt Butterfly Heist of 1947.

The Colin Wyatt Butterfly Heist is a bizarre and fascinating story, and as Braby is about to discover, it started a flutter which is still impacting science today.

The charismatic butterfly thief

Before we talk about the modern day, let's head back to 1942, when a British man named Colin Wyatt travelled to Australia to work in the Air Force, bringing his wife, Mary.

Wyatt was an Olympic champion ski jumper, a military camouflager, an author, a yodeller, a painter, a keen naturalist, and a butterfly collector. A renaissance man, if you will. In newspaper reports from the time he's also often described as 'charismatic' and 'handsome'.

Braby, in retelling this tale, prefers the term 'rogue'.

Wyatt used his charm and notoriety to gain access to Australia's most extensive museum butterfly collections. To an obsessive collector like Wyatt, a rare butterfly is as valuable as a rare diamond. So he conducted a heist – with gentlemanly flair.

Under the guise of updating a book on Australian butterflies, Wyatt was invited into the secret back rooms of museums, home to the most coveted specimens in insect collections. He then simply strolled out of the museums with little tins full of butterflies in his pockets and under his hat. On Wyatt's trip to Adelaide, it's said that he even locked himself in the museum overnight to get the job done under the cover of darkness.

Over multiple visits during 1946, Wyatt smuggled approximately three thousand butterfly specimens out of Australian museum collections. That's a lot of butterflies.

Wyatt posted the stolen collection to his home in England, and soon flew back himself. This time he was returning without Mary; their marriage had disintegrated while in Australia. Miserable in his empty house, Wyatt threw himself into relabelling all his stolen specimens with fictional collectors and locations, sometimes subbing in his own name.

Almost immediately, word spread through staff at the Australian museums that there were holes in their butterfly collections. The rarest, most difficult-to-find specimens had simply vanished.

Scotland Yard detectives were called in to investigate the case of the missing butterflies. After a year-long process, they eventually charged Wyatt, retrieved 1600 of the priceless specimens and sent them back to Australia.

Wyatt pleaded guilty to the heist, although he claimed he wasn't in his right mind after his recent divorce. The judge seemed to sympathise with this excuse. As *TIME Magazine* reported: 'The judge let him off easy (a £100 fine); he understood "the distraction of your mind" that had led Wyatt to a crime of passion.' Who knows what Mary thought.

The Australian curators were left with the job of painstakingly sorting the returned specimens back to their original archives.

An ominous yellow tag remains on each specimen to this day, saying, 'Passed through C. W. Wyatt theft collection, 1946–1947,' as a reminder that there is a small element of doubt hanging around every specimen touched by Colin Wyatt.

Revealing a fake

Skip forward 72 years, and we're back in Braby's lab at the Australian National University. While there wasn't one of those yellow tags on the Flame Hairstreak specimen in Braby's photograph, he was starting to agree with his colleague that it could be mixed up in the Wyatt case.

It fit that Wyatt, who'd exhibited his art in Australia, had the painting skills, as well as the means and motive, to forge a specimen from the Australian Museum's collection. In 1946, this was the only known example of a Flame Hairstreak in the world. There's no doubt it would have been on Wyatt's wish list.

Braby reasoned that Wyatt could have nicked the rare specimen, and then inserted the quite realistic fake he'd created in its place, so no one would suspect it was missing. But in the meticulous world of taxonomy and museum collecting, calling out a fake is an extraordinary move to make. Braby needed proof.

Besides, to finish the scientific paper they were working on –

revising the taxonomic status of this species of butterfly – Braby and Eastwood needed to examine the true original specimen, called a holotype.

A side-quest was born to get to the bottom of this mystery.

Braby travelled to Sydney to visit the museum, and to search the collection with his own eyes.

He came across one particular specimen that stood out as a very good candidate for the actual Flame Hairstreak holotype. It was in much better shape than the supposed holotype specimen he'd seen on his computer screen.

'It had a Wyatt Theft Collection label on it, a similar date in the 1940s, but it had a different location label,' Braby explains.

Here, I ask you to imagine a montage of a busy scientist zooming in on butterfly wings, investigating genetic analyses, comparing labels, illustrations and photographs, reading up on the travels of Colin Wyatt and butterfly collectors in the region, then sitting down to write copious notes.

Braby took the curator at the Australian Museum aside and explained what he'd found. He pointed out that beyond the red painted wing, there were other dodgy differences that pointed to the specimen being faked. The black bands on the hind wing of the (alleged) fake weren't the right orientation for a Flame Hairstreak, and the orange band was criss-crossed with black veins not apparent in the collector's original drawings.

In Wyatt's memoirs detailing his butterfly-collecting adventures, you can trace exactly where he travelled and when, piecing together a vivid chronology of his collecting. But Braby found that the labels of these two specimens didn't match with their collector's supposed chronologies – that is, until he switched the labels.

Then it all made sense.

Wyatt never went to the remote Barrington Tops location that the Flame Hairstreak hailed from, but he did go to the Blue Mountains, where he diarised his excitement at collecting multiple specimens of another similar species, the Silky Hairstreak (*Pseudalmenus chlorinda*) – a species that has black veins on the orange section of the wings.

'When I told the curator, he was stunned,' Braby says. 'He was just shaking his head.'

Braby and Eastwood published a paper laying out their case, concluding that Wyatt stole the original Flame Hairstreak holotype for his private collection, then replaced it with a Silky Hairstreak specimen he'd collected himself and carefully painted it as a fake, switching up the two genuine labels.

It was huge news: a fake had never been discovered in an Australian national insect collection before.

In their 2019 paper, Braby and Eastwood wrote: 'Wyatt's fraudulent and apparently unprecedented act in creating the fake holotype has gone unnoticed for 72 years and must surely rank as Australia's greatest taxonomic fraud!'

To fan the flames, after this paper came out, in 2024 yet another falsified butterfly specimen was found by a scientist called John Tennent from the London Natural History Museum. Tennent also attributed this to Wyatt's misdeeds.

The butterfly effect

Like the judge in Wyatt's trial back in 1947, some may wonder if nicking a bunch of dead butterflies is that big a deal. But this stuff really matters to science.

Wyatt stole from the public record, with no regard to how removing one-of-a-kind original specimens would derail the naming and classification of species going forward.

'It's almost beyond words, what he's done,' Braby says. 'The museum institutions are the foundation of our taxonomy and nomenclature, and hence they really underpin our knowledge of biodiversity.'

Taxonomy is the universal language which forms the basis of

all global biological systems. There's a code of conduct you must follow when describing a species that's new to science, and every step is crucial to ensure the record is accurate. The code states that when you publish a paper naming a new species, you must present a holotype, as physical proof of the species.

'The idea behind having a holotype specimen is that there's no doubt as to what the species actually looks like,' Braby explains.

Generations of scientists will refer back to the holotype again and again in their research, so the specimen needs to be publicly accessible. 'If you start tinkering with that, the fabric of taxonomy just falls apart.'

This is not to say that scientists don't make mistakes. Scientists are people – fallible like the rest of us – but they hold each other to account. An integral part of the scientific method is scrutinising the work of your peers.

So if a name or taxonomic description is published that's not in line with the code, the scientific community will almost always identify the issues and reject the invalid new species. That diligence is crucial if you want to be a good taxonomist.

But an honest mistake is very different to deliberate deceit, like Wyatt's fake butterfly specimens. Finding errors becomes a lot harder when there's no transparency, and when you can't even trust the labels of a collection. It creates lots of extra work, as Braby found on his detective detour, and it wastes time; time that's precious when there's so much to be done.

'In this age, we have two diametrically opposing issues: on the one hand, we're losing biodiversity at a phenomenal rate and on the other hand, we're discovering biodiversity on an unprecedented scale,' Braby says.

It's a strange thing to find yourself in the middle of a stark biodiversity crisis, while technology and scientific innovation is opening the realms of discovery.

'So if taxonomy is going to go leaps and bounds forward, we've got to get it right,' Braby says.

The heist might be history, but the future demands we put it right.

'We've got to learn from our mistakes and try not to repeat these things.'

✳ *Cat-astrophe: Australia's feral cat problem*, p. **33**
Alternative accommodation, p. **213**

STANFORD PRISON EXPERIMENT: WHY THE LEAD PSYCHOLOGIST DEFENDED HIS INFAMOUS STUDY TO THE END

Rachel Fieldhouse

A world-first shyness clinic, an analysis of sexist *Reader's Digest* jokes and perhaps the most infamous study ever all emerged from one man's mind: Emeritus Professor Philip Zimbardo.

The psychologist died in October 2024 at his home in San Francisco, California, aged 91.

He wrote hundreds of papers spanning cults, hypnosis, altruism and the psychology of evil.

In 1974, he concluded that most of 1000 *Reader's Digest* jokes were anti-woman, demonstrating how non-conscious sexist beliefs were maintained.

He also led the notorious Stanford prison experiment, which as he eventually acknowledged, made its findings 'at the expense of human suffering'.

In the decades after World War II, various researchers used field studies to examine the nature of extremism and evil. Professor Zimbardo received funding from the US Office of Naval Research to recruit 21 healthy male college students to stay in a mock prison in the basement of Stanford University's psychology department.

The mostly middle-class and white students received US$15 a day (A$181 in 2024) to act as guards or prisoners – randomly assigned – for two weeks.

The experiment lasted six days.

As Professor Zimbardo's team wrote in the *International Journal of Criminology and Penology*: 'We witnessed a sample of normal, healthy American college students fractionate into a group of prison guards who seemed to derive pleasure from insulting, threatening humiliating and dehumanising their peers.'

Professor Zimbardo's conclusion was that people were not cruel because they were 'bad apples'. They were influenced by their environment – in this example, an environment he had created.

The experiment

The three 1.8-metre by 2.7-metre cells sat within a 10-metres-squared cell block, with separate guards' quarters and a closet for solitary confinement, to be used for a maximum of one hour per offence.

The guards lived 'normal lives' outside of their eight-hour 'shifts', but the prisoners stayed 24 hours a day.

On day one, real Palo Alto City police officers 'arrested' the prisoners and handed them to a researcher, who blindfolded them and drove them to the mock prison. The prisoners were stripped naked, sprayed with deodorant, given a uniform and escorted to a cell.

Each day, they received three bland meals and had three toilet visits under supervision from the 'guards'.

Guards performed headcounts at least twice per day, sometimes at 2 am.

The outcome

On day two, the prisoners barricaded themselves inside the cells.

The guards responded by blasting the prisoners with fire extinguishers, breaking through the barricades and forcing the ringleaders into solitary confinement.

On day three, the guards forced prisoners to defecate in buckets rather than toilets and refused to feed 'bad' prisoners.

On day four, one prisoner stopped eating and was put in solitary confinement for three hours – the limit should have been one hour.

By day five, half of the prisoners had left the study due to 'extreme emotional depression, crying, rage and acute anxiety', or in one case, a body-wide 'psychosomatic rash'.

While this happened, Professor Zimbardo acted as prison 'superintendent', advising the guards and meeting with prisoners. He later told university alumni publication *Stanford Magazine* that by day three 'I had become the superintendent of the Stanford county jail ... not a researcher at all'.

The researchers halted the study on day six because of the 'unexpectedly intense reactions'.

But Professor Zimbardo's wife of 52 years said it unfolded a little differently.

Professor Christina Maslach – known for developing the Maslach Burnout Inventory – was one of Professor Zimbardo's graduate students at the time. She told *Stanford Magazine* that she saw the guards chain together prisoners' feet and put bags on their heads when bringing them out for a toilet break.

This prompted her to confront Professor Zimbardo. 'I feared that if the study went on, he would become someone I no longer cared for, no longer loved, no longer respected,' she said.

The pair married a year later, in 1972, and remained together until Professor Zimbardo's death.

The ethical breach

The study had ethics approval from the Stanford Human Subjects Review Committee, but its early termination led to stricter guidelines for human experiments.

It spurred the National Commission for the Protection

of Human Subjects of Biomedical and Behavioral Research to produce the 1978 Belmont Report, which codified three principles for human participation.

These were respect for persons (ensuring that people entering research do so voluntarily and with enough information), that people are treated ethically and protected from harm, and that the benefits of research are fairly distributed.

The extent to which Professor Zimbardo's team provided information – in line with the first principle – continued to stir debate.

In 2019, a researcher from France's Côte d'Azur University, Dr Thibault Le Texier (PhD), accused Professor Zimbardo of not really running an experiment, but an exhibition of his pre-existing belief that prisons were toxic.

'Instead of trying to neutralise the potential effects of this conviction on his objectivity, Zimbardo designed his experiment from the outset as a demonstration of the toxicity of prison,' Dr Le Texier wrote in *American Psychologist*. He said participants working as guards knew Professor Zimbardo wanted them to act aggressively and 'hammed it up'.

As one 'guard', Dave Eshelman, said in 2011: 'I set out with a definite plan in mind, to try to force the action, force something to happen, so that the researchers would have something to work with.'

Another prison experiment?

Dr Le Texier also argued that the experiment's findings were worthless due to selective sampling, as they were based on 21 hours of video and audio from the 150-hour study.

In 2001, UK researchers Professor Alex Haslam and Professor Steve Reicher declared that trying to reproduce the results in the 21st century was ethically 'impossible'. 'One of the great problems arising from the Stanford prison experiment was that it made strong claims ... but then led to debate being closed off because further research was declared to be ethically unacceptable,' they wrote in the *British Journal of Social Psychology* in 2001.

Then came their 'BBC prison study'. The two UK researchers created an environment resembling 'hierarchical institutions' such as prisons, schools and barracks. They randomised 15 men to be guards or prisoners and employed daily psychometric testing.

However, their experiment did not produce the 'tyranny' of Stanford, with the BBC prison study guards struggling to unify when the prisoners collectively challenged them.

'The results of the BBC prison study suggest that the way in which members of strong groups behave depends upon the norms and values associated with their specific social identity and may be either anti- or prosocial,' Professors Haslam and Reicher said.

Perhaps being filmed for the UK's public broadcaster was another factor.

Professor Zimbardo never changed his view that his infamous experiment deserved respect for its findings.

But he eventually acknowledged that the study was unethical, even though no participants reported long-lasting trauma.

In his 2007 book *The Lucifer Effect* he said: 'I was guilty of the sin of omission – the evil of inaction – of not providing adequate oversight when it was required. The findings came at the expense of human suffering.

'I am sorry for that and to this day apologise for contributing to this inhumanity.'

✱ *Is this actually PTSD? Clinicians divided over redefining borderline personality disorder*, p. **63**
Some psychedelic medicine developers want to ditch the therapy aspect. What could go wrong?, p. **120**

DEEP TIME ENCOUNTERS IN THE GARDEN

Fiona McMillan-Webster

To enter a garden is to become an explorer of deep time. There are gardens that have been tended for years, for decades, for centuries, but even the newest gardens are filled with ancient stories. A recently dug vegetable patch can tell a prehistoric tale, and the most fiercely tended rose gardens contain something primaeval. In fact, if you take a closer look in any garden, you can find stories that will take you back in time many millions of years.

Enter a garden and seek out the dark, shaded places. There you might see tiny liverworts straining out of the soil. Or perhaps you'll find patches of moss clinging to stones or the base of a tree. These are bryophytes and they are among the closest living relatives to the first plants on Earth, which began to evolve from photosynthetic green algae around half-a-billion years ago. Mainly aquatic in nature, the earliest plants had lingered in the shallows for aeons, slowly contriving a means of anchoring in place via primitive root-like structures called rhizoids. Find a patch of moss in a garden and notice how it still uses these fine, hair-like protrusions to latch onto surfaces. These rhizoids are often just a fraction of a millimetre wide, but what they lack in size they make up in numbers and their collective strength is such that a patch of moss adhered to a stone can be quite difficult to budge. This tenacity is a remnant of the moment plants first moved onto land and resolved to stay there.

At the time, the oceans were teeming with a staggering diversity of life forms thanks to the Cambrian explosion, yet land was still a barren, hostile place. It was dry, of course, and access to nutrients and water was wildly unreliable, never mind the lethal levels of UV and extreme daily and seasonal temperature variations. With the help of mycorrhizal fungi, which facilitated access to water and nutrients, the first land plants began to green the edges of streams, rivers and freshwater lakes, kick-starting a huge rise in levels of atmospheric oxygen and enticing a few small members of the animal kingdom to follow them onto land. The world has never been the same since.

Any mosses and liverworts you encounter in a garden may seem delicate, but they are tougher than they look and have retained survival skills from their earliest days, such as the ability to tolerate periods of extreme dryness. You may also notice that modern bryophytes are quite small. Most lack the structural integrity to support their own weight beyond a few centimetres and, although they contain simple conductive cells, they rely heavily on diffusion to distribute water and nutrients. Perhaps the entire plant kingdom would have remained tiny and a bit fuzzy if some plants hadn't figured out an interesting workaround.

An easy thing to find in any garden is a branch. They're certainly everywhere. If you were asked to count the branches on just one modestly sized tree, it might take a few minutes. Most of us would begin by counting the number of places a bough emerges from the main trunk, and maybe any robust-looking stems emerging from these. Yet even so, we wouldn't come anywhere close to counting the number of places branching occurs on that tree. Think of each stem, each leaf, each bifurcating vein, each divergence of a microscopic venule – it would take years to count every branching event on that tree. The reason for this takes us back more than 440 million years and has quite a lot to do with why we even have trees at all.

Back then, around the beginning of the Silurian period, a few ancestral bryophytes began making complex structural molecules, such as lignin. These biopolymers added rigidity to cell walls, allowing the formation of strong, water-impermeable vasculature tissue, which allowed them to channel water and nutrients. Importantly, this vasculature was able to branch. Vessels could divide and divide again, allowing plants to maximise their surface area. These stronger, well-plumbed plants not only grew taller, but deeper as well. They developed true roots, which afforded greater stability as well and an opportunity to mine downwards into the Earth for minerals and water. By the time the Silurian gave way to the Devonian 419 million years ago, vascular plants were on the rise, producing more oxygen and drawing down more CO_2, and they were only just getting started.

Today, every leaf you see is like a postcard from the Devonian. Look closely and you'll find they are crowded with vascular tissue, whether it's fine striations of veins diverging from a main stem or pale capillary nets branching this way and that. With the help of branching vasculature, botanical diversity exploded in the Devonian. It became a time of youthful exuberance for leaf-bearing plants – their salad days, you could say. The first ferns arrived and then the first true trees, with robust, woody trunks reaching heights of 20 to 30 metres, and branches covered in fern-like foliage. Yet they all remained bound to the water's edge, unable to venture much further than that.

In every garden you will find seeds, multitudes of them. To pick up just one seed and cup it in the palm of your hand is to hold something truly revolutionary. Despite all the progress made in the first half of the Devonian, dry land remained a forbidding place for plants. As is still the case with their modern counterparts, early bryophytes and ferns – and even those first trees – reproduced via spores, which required the presence of surface water. A single plant might release vast numbers of uniformly sized spores, but only those that came into contact with moisture could germinate into a gametophyte, which is an intermediate generation that produces sex cells. The crucial next step – fertilisation – required water, as it enabled the male cells to swim to the female cells.

Over the course of millions of years, something shifted. Some plants began releasing smaller microspores and larger megaspores, which matured to produce male and female sex cells respectively. As time passed, these plants began to hold on to their megaspores, protecting them in cup-shaped structures, called 'cupules', as they matured. These cupules also enabled plants to catch wind-blown microspores, bringing them into contact with the female cells. Plants no longer needed an open body of water for reproduction – not a stream, a lake or even a puddle. Fertilisation took place right there on the plant, and so did the development of the tiny plant embryo. The first seeds had arrived, and they were about to change the world yet again.

All gardens quietly teem with life, but they are also places of still-ness. Arguably, this effect is mostly due to the plants. They might sway in the breeze, but they don't scurry away as you approach or roam around in search of more sunlight or a drink. Rooted in place by tiny rhizoids or sturdy roots, plants are generally immobile. This is fine if conditions are good, but disastrous if conditions change for the worse. For any plant population to survive changing conditions or to simply establish new niches, the next generation must be able to travel. Seeds opened up new territories for plants to occupy, whether a few metres away from a riverbank or hundreds of kilometres inland – or, if the winds were favourable, some distant island.

Seeds not only enabled plants to disperse their progeny across physical space, they could also send them forward in time. It's tempting to think of a seed as a hard mishmash of plant cells, but the overwhelming majority of seeds contain a tiny immature plant, complete with a stem, a root structure and at least one embryonic leaf. As the first seed-bearing plants nurtured their growing embryos, they began enclosing them in a seed coat and an outer layer of protection before releasing them into the harsh world when the time was right. We see this story in our modern gymnosperms, which includes conifers, cycads and ginkgo. Indeed, a pinecone tells a tale of staunchly protective plant parents over 300 million years ago.

Eventually, plants began to provide a packed lunch as well. Recent evidence suggests that the first flowers appeared at least 164 million years ago during the Jurassic. They were rare, tiny things – most smaller than a millimetre – but by the time the Cretaceous was underway, flowers were getting larger and blooming everywhere. The first ancestral magnolias arrived, as did the earliest roses. Flowering plants began to fill their seeds with endosperm – a supply of carbohydrates and nutrients that surrounds the baby plant and supports its growth until the day it can push out its nascent leaves and start making its own food. Seed coats often became tougher, too, allowing embryos to quietly endure bad conditions, sometimes for years. In essence, plants improved the odds for their offspring's survival by investing resources and offering protection, buying them time until the right conditions for growth appeared. Fruit encouraged travel. Some plants grew sweet fruits around their seeds to entice animal dispersers; others opted for harder, spiky fruits to catch on passers-by; still others formed papery-thin aerodynamic fruits to float on the wind. To sit in a garden and watch a seed drift on the air or admire a magnolia or take a bite of an apple is to experience a little bit of the Cretaceous.

Deep-time stories like these are not only found in gardens, of course. For those who can venture further afield than cities and their sprawling surrounds, there are forests to explore, open tundras or wild meadows, and plant-covered cliffs overhanging the sea. Yet nearly 60 per cent of the global population now lives in urban centres and this figure is set to grow starkly in the coming years. Increasingly, it is in built places full of concrete, steel, bitumen, glass and plastic that our lives play out day after day and year after year. In the midst of all this, gardens offer a tangible point of connection with biodiversity. So, too, does every stretch of parkland, every rooftop vegetable patch, every wild verge, and every micro-forest allowed to occupy a few square metres in the centre of a carpark. These green spaces are where ecosystems can emerge, thrive and interconnect. In a sense, they are also thin places, where boundaries waver and time contracts, and for a short while we are brought close to distant pasts and possible futures.

✳   *'Earth poetry' in the Arctic*, p. **53**
    A freediver finds belonging without breath, p. **69**
    Sounds of the slow-rolling sea, p. **175**

THE UNEXPECTED POETRY OF PHD ACKNOWLEDGEMENTS

Tabitha Carvan

'Embarking on a PhD is a treacherous task for even the most bold and brave of this world, and yet here I sit, writing my way through the final hoops of this life-changing experience.'

James Beattie, *The statistics of magnetised interstellar turbulence* (2024)

For a reason I can no longer remember, I started picking science theses at random from the Australian National University library catalogue and reading only the acknowledgements.

'The acknowledgements are my favourite part of all my thesis.'

Nian Jiang, *Growth and characterisation of GaAs/AlGaAs core-shell nanowires for optoelectronic device applications* (2016)

'This part of the thesis is the one that I was most looking forward to write.'

Jorin Diemer, *A mathematical model of ion homeostasis in the malaria parasite,* Plasmodium falciparum (2022)

Once I started, it was very hard to stop.

Sometimes I found myself reading them at home, on my own time, always thinking, 'Just one more.'

'If you, the reader, are a PhD candidate, you should know that not everyone's doctoral experience is the same, but I found mine quite challenging.'

Zac Cranko, *An analytic approach to the structure and composition of general learning problems* (2021)

'I don't think it's overstating events to say that at the end of my PhD I find myself a rather different person than when I began (and not only because of the ravages of time).'

Elizabeth Krebs, *Breeding biology and parental care of the crimson rosella* (1999)

'I have always thought to some degree, my supervisors really took a gamble on me. I might have turned out to be completely hopeless. At some points near the start of my candidature I'm sure they thought I was.'

Andrew Casey, *A tale of tidal tails in the Milky Way* (2013)

All the acknowledgements followed the same basic structure; they're formulaic. But as with many formulaic things, there's a story behind each one.

I read hundreds and hundreds of acknowledgements.

'The production of a dissertation is a formidable, arduous and demoralising task.'

Mahyar Bokaeeyan, *Analytical and approximate methods in rogue wave theory* (2020)

I came to see that the acknowledgements of a PhD thesis are their own kind of thing.

The rest of the thesis contains careful, reasoned findings and figures, but on this one page, the author–scientist can release all the pent-up emotion they couldn't express elsewhere.

They're like an explosion in a lab.

'When looking at the acknowledgements of others, I have found it is common practice for scholars to reserve their purplest prose and most overblown sentiment for their acknowledgements. If I had the skills, I would do so myself.'

Martin Worthy, *A history of fire and sediment transport in the Cotter River catchment, southeastern Australia* (2013)

'I have to laugh as I recall certain fatigue-driven moments of internal melodrama, usually after long and unbroken spells of intense laboratory work, wondering how on earth I was going to singlehandedly manage this project to completion.'

Jenna Roberts, *A multi-disciplinary assessment of endocrine disrupting chemicals, pharmaceuticals and personal care products in Australia's largest inland sewage treatment plant and the Molonglo/Murrumbidgee effluent-receiving environment* (2015)

The acknowledgements have a quality which is hard to describe.

They feel like they've been drafted a hundred times in the head of the author, but then put down on the page in a hurry, the clock ticking on their deadline.

Like, they're trying to tell you the most important thing they've ever said – at the very moment the ship is pulling away from the dock.

'Tibor and Jena, Elisa and James, how can I express how much all your love and support has meant to me?'

Allie Mokany, *Resilience and resistance of ephemeral aquatic ecosystems to environmental change* (2006)

'Finally I thank you Kat, for being the only thing that really matters.'

Jevon Longdell, *Quantum information processing in rare Earth ion doped insulators* (2003)

The urgency of the task, combined with its enormity, makes it seem almost impossible to complete.

'Truly, I could never find the words to thank you for your support. There are so many of them and they catch in my throat even just thinking of them, and make it ache.'

Inez Harker-Schuch, *Using 3D serious gaming interventions to promote climate science literacy in the 12–13-year age group* (2021)

'At this point, to describe the thankfulness I want to express only seems to make it seem smaller than it really is, because words can only do so much.'

Matthew Crabb, *Nonlinear wave patterns in the complex KdV and nonlinear Schrödinger equations* (2022)

Many people find themselves lost for words at this critical moment.

It's frustrating for scientists to not be able to accurately represent their data.

'For all that Tim has done to shape me into the person and the scientist that I am today, "thank you" seems so embarrassingly simple and utterly inadequate to convey the depth of my gratitude to him. I truly hope I have done him proud.'

Ponlawat Tayati, *Molecular wet adhesion* (2015)

'I cannot begin to describe their level of kindness and generosity for which they – being the good souls they are – have expected nothing in return.'

Ryan Kirk, *Synthesis and coordination chemistry of the arsacyclopentadiene ligand* (2023)

'In some ways, this section of my thesis is the hardest to write, and also the most important.'

Sandra Ann Binning, *The effects of biotic and abiotic factors on fish swimming performance* (2014)

But even an English literature student would surely struggle with this writing project. Within this generic format, you need to deliver an outpouring of love and gratitude on the scale of a wedding vow.

And then throw in the in-jokes of a 21st birthday speech, the teary-eyed reflections of a eulogy, and the celebration of a birth announcement.

'It takes a village to raise a child and this PhD is as close to bearing a child as I'll likely ever come.'

Rachael Gross, *At a crossroads: African elephant conservation, climate change and community-based management* (2023)

'He has had to live with me and the emotions that go with a PhD and has stood steadfastly by me encouraging me to the bitter end, though I'm sure he would like to know just when the demonic banshee moved in and I checked out.'

Trudi Wharton, *Biology and ecology of* Essigella californica *(Essig) on* Pinus radiata D.Don. *in Australia* (2005)

'There are many people who come in and out of your life and who make a difference. There are deaths and births; there is despair and joy.'

Jennifer Metcalfe, *Rethinking science communication models in practice* (2019)

'Thanks also to Moira, Zoë and Brad for tolerating my existence.'

Michael Duglosch, *Cross-coupling chemistry as a tool for the synthesis of diverse heterocyclic systems and natural products* (2019)

'Tom, few supervisors can say that their students have nearly killed them and mean it literally. Thanks for everything.'

Iain McConnell, *Substrate interactions in the Photosystem II water oxidising complex* (2008)

But in this, the most unscientific part of a PhD submission, you also see how the science is done.

No matter how impenetrable the thesis title, the project's success always seems to come down to the same simple thing: other people.

'Thank you to those individuals who donated their own teeth to my study, and to the late great Professor Colin Groves, who eagerly donated a dugong tooth for a laser ablation Sr standard, and then eagerly donated a second one when the first one went missing.'

Hannah James, *Adventures in archaeological science: An exploration of oxygen and strontium isotopic variability on a micro- to macro-scale* (2021)

'To Banak Gamui, I thank him for purchasing a roll of fishing line and a pair of gloves for the parrot project, and for which I never paid his money back.'

Paul Igag, *The conservation of large rainforest parrots: A study of the breeding biology of palm cockatoos, eclectus parrots and vulturina parrots* (2002)

These are advanced research projects, using high-tech equipment and sophisticated analysis, and yet the thing which is acknowledged, again and again, as if it's of equal, if not greater, importance is conversation.

Just, you know, talking.

'I would like to thank Dick Henley for his unrelenting enthusiasm for volcanoes and magmatic gases.'

Dominique Tanner, *In situ mineral geochemistry as a guide to ore-forming processes* (2014)

'I am thankful for Jim's patience (especially with my typos), his discussions, his time, and his open-mindedness to my sometimes unusual ideas.'

Larissa Huston, *The high pressure phase transformations of silicon and germanium at the nanoscale* (2019)

'Laurence was very supportive about my interest in thrips even before I had a thrips project, and I will never forget some of the conversations we had in his office.'

Brian Garms, *Native insects as a framework for understanding potential impacts of exotic species* (2014)

'Thanks for letting me barge into your office whenever I came across problems with the micro-PL system and simply answering "I'll be right there".'

Nian Jiang, *Growth and characterisation of GaAs/AlGaAs core-shell nanowires for optoelectronic device applications* (2016)

'Thank you for patiently listening to my long conversations on Sonogashira reactions and sharing my frustrations when I couldn't grow crystals.'

Chriso Maria Thomas, *Construction of porous supramolecular frameworks assembled from covalent cage compounds* (2021)

'Sometimes just having someone to listen to me when my research was at a stumbling block was all that was needed to find a solution, and I thank David for his patience and time in this regard.'

Janet Pritchard, *Linking fish growth and climate across modern space and through evolutionary time: Otolith chronologies of the Australian freshwater fish, golden perch* (Macquaria, ambigua, *Percichthyidae*) (2004)

'Thank you to the friends who simply knew not to ask.'

Sarah Tynan, *Interpreting environmental change using bivalve shell geochemistry* (2017)

After a while, I stopped dipping into the theses from before the 1980s.

Back then, acknowledgements were a businesslike thank you to supervisors, collaborators, funding bodies and typists.

Still, I'm sure there are stories in there too.

'Finally, I thank Mrs Barbara Geary, both for her excellent typing and also for her willingness to accept, and ability to decipher, the manuscript drafts of the thesis.'

John Groves, *Varieties of soluble groups* (1971)

'Mrs Barbara Geary for her work in typing this thesis.'

Yu Kiang Leong, *The CREAM Conjecture and certain abelian-by-nilpotent varieties* (1972)

'For her excellent typing I thank Barbara Geary.'

Langdon Harris, *Varieties and section closed classes of groups* (1973)

'I particularly want to thank Mrs Barbara Geary for her excellent typing of this thesis. She has endured many awkward changes and revisions without complaint, and has transformed my messy drafts and redrafts into a beautifully typed manuscript.'

Abul Kasem Muhammad Masood-ul-Alam, *The topology of asymptotically Euclidean static perfect fluid space-time* (1985)

Now, there are no limits to who, or what, you can thank.

'Finally I wish to acknowledge the work of Dave Grohl who provided a fine soundtrack for much of this PhD study.'

Andrew Sullivan, *Competitive thermokinetics and non-linear bushfire behaviour* (2007)

'I would also like to thank the various astronomers who forgot to collect their desserts from the fridge over the years (they were not wasted).'

Roberto Soria, *Accretion processes in black-hole binaries* (1999)

'Thanks to my dog Tonks for the joy and cuddles.'

Catherine Ross, *Bringing back the bettong: Reintroducing ecosystem engineers for restoration in Box-Gum grassy woodland* (2020)

'Shout out to my friends on and off Stromlo, for their patrician taste in hot pot and KFC.'

Jane Lin, *Galactic archaeology: The Milky Way in the context of large scale surveys* (2022)

One of the most-repeated phrases in the acknowledgements is: 'I could not have done it without you.'
Every time I read it, I believed it to be 100 per cent true.

'Finally, thank you so much to my amazing family for their love, especially my wife, Maansi Joshi. You are my shining light, my daily inspiration and I could not have done this without you.'

Geoffrey Kay, *Scaling the benefits of agri-environment schemes for biodiversity conservation in agricultural landscapes* (2016)

'Finally, to Mum, your courage is inspiring. You may think I am the rock in the family, but I could not have finished my PhD without you.'

Samuel Drew, *Explorations in polyene chemistry* (2015)

'And finally to my beautiful wife, Jing, I could not have done this without you and I cannot wait to spend the rest of my life with you.'

Roger Coulston, *Cyclodextrin nanomachines at work* (2009)

A PhD takes at least three years, and often many more, during which life necessarily goes on around you, or without you.
The cost is very real.
If you are in the lab, then you are not at home with your family. If you have relocated for your studies, you won't be having home-cooked meals.

'Six years is a long time.'

Philip Palma, *Laser-induced fluorescence imaging in free-piston shock tunnels* (1999)

'As their only child, my parents never stopped me from pursuing my dream. They unconditionally supported me to continue my study in Australia. Even if it means that I have to be thousands of miles away from them'.

Kun Peng, *III-V compound semiconductor nanowire terahertz detectors* (2016)

'And thank you for picking up the slack with house chores whenever I was bogged down with lab work – I know I was the reason you bought that Roomba.'

Lauren Harrison, *Sex and conflict: How competition shapes reproduction, behaviour and life-histories in various animals* (2022)

And sometimes, life will simply come at you, whether you want it to or not.

'First, I want to acknowledge that the last year of my PhD, 2020, has been exceptionally hard for everyone. I am incredibly grateful for being safe and healthy.'

Raktima Dey, *Understanding historical and future changes in mean and extreme rainfall in Australia* (2020)

'I am grateful to all the persons who have helped me recover from the fire that devastated the observatory and destroyed my house at Mount Stromlo (at a very delicate moment of my thesis too ...). I will remember everyone who shared those days with me, so full of tragedy but also of hope and all the best human feelings.'

Maria Salvo, *How good are type Ia supernovae as distance indicators?* (2020)

'Finally, to our baby, "Bubbles", whose gentle kicks have reminded me I haven't been alone in the final months of this thesis. But, Bubbles, if you can just sit tight for a few more weeks so that Mum and Dad can have a rest we'd be very grateful.'

Kelli Gowland, *Investigations into the phorophyte and fungal relationships of three epiphytic Aeridinae orchid species in southeastern Australia* (2008)

I'm sure there are detractors behind many a thesis, but in the acknowledgements, they only exist by omission.

And if the whole experience was an ordeal? Thankfully that's now in the past.

Here, on this page, it's a perfect world. There's no one to blame and everyone to thank – including baristas.

'Doing this thesis required considerable amounts of coffee. I want to recognise Luke, our Barista in the Little Pickle Café, for the excellent coffee and the funny attempts to learn Spanish.'

Claudia Múnera Roldán, *Narratives of adaptation for future-oriented conservation* (2017)

'Mum, you once told me I could do whatever I set my mind to, I told you that was a cliché, and you said, "Yeah but for you it's actually true." Well here you go, look at the trouble that attitude has led to.'

Thomas Loan, *Cell lysate as a platform technology for biocatalytic synthesis and nucleic acid amplification* (2020)

'And of course to my parents, Andrew and Margaret; my foundation, my teachers, my wings.'

David Blair, *Comparisons of vegetation recovery post-fire, logging and salvage logging in the Victorian Central Highlands* (2019)

'Finally, to my parents: this may have all started when we visited the Sydney Observatory one birthday and saw the marks of Shoemaker-Levy 9 just spinning into view on the disk of Jupiter, but it was your love and support all these years since that made it possible.'

Michele Bannister, *Bright trans-Neptunian objects in the southern sky* (2014)

'This thesis happened because my family and good friends believe in me.'

Jo Leen Lim, *Protein engineering of* Escherichia coli *β-glucuronidase* (2017)

'I thank my mother, Dr K. Lalitha for being the person I aspire to be. I wish I could be half person you were. Words can't express how much I miss you ...'

Mayuri Sathyanarayana Rao, *On the detection of spectral distortions in the CMB: Recombination to reionisation* (2017)

Why wouldn't you want to read 'just one more'?

'You are all precious like stardust.'

Dominik Koll, *A 10-million year time profile of interstellar influx to Earth mapped through supernova Fe-60 and r-process Pu-244* (2023)

Who wouldn't want to extend their time in such a world just a little bit longer?

'This small family, including our cats, Molly and Jesse, is my true treasure.'

Xilin Lu, *Light-matter interaction models: Symmetry and non-hermiticity* (2023)

'And to Louise, thank you for everything. I don't know how you do it.'

James Hennessy, *Modifying enzyme catalytic pathways* (2009)

I wondered how many of these acknowledged loves endured, not to mention how many scientific or academic careers.
But for now, on that page, the author knows nothing of what the future will hold.

'So long and thanks for all the coffee!'

Samantha Burgess, *Geochemical ecology of temperate corals* (2007)

They are frozen in this moment in time between an ending and a beginning.

'YES!! After a looooooong journey, I've finally done it! YES!!'

Yi-chi (Candace) Tsai, *Oxidative behaviour of O-methyltyrosine and p-methoxyphenylglycine derivatives* (2007)

It is the best moment.

'It seemed as though the day that I would type these acknowledgements would never come.'

Amy Constable, *The emergence, formalisation and evolution of biodiversity offset use in Australia* (2022)

'But at this point, looking back over the mountains you climbed and the valleys you crossed, you realise your motivation came from the people around you. That is why we have an acknowledgements section.'

Liam D Bailey, *Between the devil and the deep blue sea: Consequences of extreme climatic events in the Eurasian oystercatcher* (Haematopus ostralegus) (2016)

It's the moment when it's done.

'I am overwhelmed. I have finally completed this milestone, three and a half years of my life culminating in this body of work.'

Andrew Shafik, *5-methylcytidine has a complex, context-dependent role in RNA* (2017)

'It will take me the rest of my life to repay you for your patience, support, encouragement, dedication, and sacrifice.'

Richard Turner, *Avian life histories in a changing world: Combining remote sensing with long-term monitoring of the superb fairy-wren* Malurus cyaneus *in Australia* (2023)

'Thank you, again and again.'

Yi-Yang Chen, *The nature and significance of macroalgae-epifauna-invertivorous fish trophic links within a macroalgal-dominated reef ecosystem* (2022)

'I'm done.'

John Dawson, *Satellite radar interferometry with application to the observation of surface deformation in Australia* (2008)

CONTRIBUTORS

MATTHEW WARD AGIUS is a journalist for Germany's international broadcaster DW. His work has been published in *Cosmos Magazine*, Australian Community Media mastheads, the *Advertiser, InDaily* and *The Wire Current Affairs*.

ROBYN ARIANRHOD has a PhD in general relativity and is an affiliate in Monash University's School of Mathematics. Her articles have appeared in *Cosmos Magazine*, *The Conversation*, *Australian Book Review* and elsewhere, while her books have been shortlisted for national literary awards and translated into several languages.

ANTHEA BATSAKIS (she/her) is an award-winning journalist, editor and content writer from Melbourne, Australia. Her work spans the environment, science, politics and the arts, but she's particularly fond of spotlighting species forgotten by conservation. When she's not freelancing, she works as the content editor at Victorian Opera.

JACINTA BOWLER has spent almost a decade as a science journalist, mostly asking very smart people dumb questions. They are a science reporter at the ABC, and have previously worked with *Cosmos Magazine*, *Nature* and ScienceAlert.

TABITHA CARVAN is a writer based in Canberra. Her work has been featured in publications including the *Guardian*, the *Sydney Morning Herald* and *The Age*, *The Saturday Paper*, *Galah*, *Australian Geographic* and three previous editions of *The Best Australian Science Writing*. She is also the author of the memoir *This Is Not a Book about Benedict Cumberbatch* (Fourth Estate).

ANNE CASEY, the author of six poetry collections, is widely published internationally, ranking in the *Irish Times*' Most Read. Her awards include the American Writers Review Prize, AAALS Poetry Prize and Henry Lawson Prize. She holds a PhD from the University of Technology Sydney, where she teaches creative writing.

JO CHANDLER is an award-winning freelance journalist, author, editor and journalism educator at the University of Melbourne. Her focus is on explanatory reporting, exploring a diverse range of issues including the science and impacts of global heating, with a particular focus on the Pacific.

OLIVIA CONGDON is senior science writer for the Australian National University and is a self-confessed nature nerd. She was previously a science communicator with the Australian Academy of Science. Olivia's work has appeared in publications including *Australian Geographic, BBC Wildlife* magazine, *Galah* magazine, the ABC, *Weekend Birder* and *Cosmos Magazine*.

GEMMA CONROY is a freelance science journalist from Sydney who is currently based in Mexico City. Her work has appeared in the *New York Times, Scientific American, ScienceAlert, New Scientist, Smithsonian* and other outlets. She has also worked as a staff reporter for ABC Science and *Nature*.

OWEN CUMMING is a freelance science writer with a masters degree in Science Communication. His work explores how we relate to science (be it a process, an institution, or a concept), telling stories that weave complex topics into the context of people's lives.

ANGUS DALTON is the science reporter for the *Sydney Morning Herald*. He was the co-founding editor of *Sweaty City*, a magazine about climate change and urban ecology, and his work has also appeared in *Australian Geographic*, *Overland* and *Voiceworks*.

ZOWIE DOUGLAS-KINGHORN is a Tasmanian writer with recent work in *Meanjin*, *Island*, *Overland* and elsewhere. Her writing has won the Scribe Nonfiction Prize, the Ultimo Prize, and been studied for high school exams. She is an Emerging Critic with the *Sydney Review of Books* and the previous editor of *Voiceworks*.

CERIDWEN DOVEY writes fiction (*Only the Astronauts*; *Only the Animals*; *Mothertongues*; *Blood Kin*) and non-fiction (*On J.M. Coetzee*). She's the recipient of an Australian Museum Eureka Award for her science writing, and co-founder of the Archival Futures Film Collective. Ceridwen is currently a Macquarie University Research Fellow and a Powerhouse Artistic Associate.

LUCINDA DUXBURY is part PhD student, part accidental poet, and part very excited, very passionate moving ball of chaos. Her writing is inspired by a world filled with wonder, novelty and the crushing weight of the climate crisis. She loves her friends and is scared of breaking expensive lab equipment.

RACHEL FIELDHOUSE is a Sydney-based journalist for *Nature* who is fascinated by the quirky side of science. She's worked as a reporter for *Australian Doctor* and *OverSixty*, and has written for the *Medical Republic*, *The Local Rag* and *Concrete Playground*. She is still banned from talking about worms at home.

RICH HARIDY is a science journalist based in Melbourne. His work has recently focused on the prolific world of psychedelic science. Rich's writing has been featured in *Nature*, *Cosmos Magazine*, *Salon*, *New Atlas* and *The Best Australian Science Writing 2024*. He is currently working on a book delving into the past and present of psychedelic research in Australia.

AMALYAH HART is a freelance science writer based in Melbourne. Her work runs the gamut from energy and climate to consciousness and AI, and has been featured in *The Saturday Paper*, *Cosmos Magazine* and others.

ANGELA HEATHCOTE is a journalist and documentary filmmaker with a focus on the environment.

MARK HORSTMAN is a science journalist specialising in polar research, currently with the Australian Antarctic Program Partnership at the University of Tasmania. In another life he worked as a broadcaster and documentary maker for Radio National and ABC TV (*Catalyst*). Mark is also a proud co-founder of *Tasmanian Inquirer*, an online news site.

PETER DE KRUIJFF is an award-winning reporter who writes about the environment for ABC Science. He previously worked at metropolitan and regional newspapers all around Australia. He is based in Boorloo-Perth on the banks of the Derbal Yaragan.

MICHAEL LEACH is an academic at Monash Rural Health and an award-winning poet. Michael's poems have appeared in journals such as *Cordite*, anthologies such as *The Best Australian Science Writing 2024*, and his books *Chronicity*, *Natural Philosophies* and *Rural Ecologies*. He lives in Bendigo on unceded Dja Dja Wurrung Country.

DYANI LEWIS is an award-winning science journalist based in Melbourne. She is a regular contributor to *Nature*, where she reports on evolution, the environment, health, and science's impact on public policy. Her work has been published in *The Monthly*, *Science*, *Cosmos Magazine*, the *Guardian*, the *Atlantic* (via *Undark*) and elsewhere.

TYNE LOGAN is the weather reporter for the ABC's Climate Team, covering stories of how weather and climate impact daily life, from extreme events to long-term climate change. She is a multi-platform reporter and presenter, having worked across print, radio, TV and podcasting.

DONNA LU is a journalist at *Guardian Australia*. She was previously a reporter for *New Scientist* magazine in London. She edited *The Best Australian Science Writing 2023* and has also had work anthologised in the 2020 and 2021 editions.

SMRITI MALLAPATY is a senior reporter for *Nature* magazine, based in Sydney, Australia. She covers the Asia-Pacific region and reports on infectious diseases and biological sciences, among other subjects. Before joining *Nature*, she worked as a freelancer in Kathmandu, Nepal.

NATASHA MAY is a journalist with the *Guardian* and covers health news in Australia. She studied English literature at university, focusing on the novel's representations of consciousness and the workings of other minds. As a health reporter, she enjoys learning more about the diversity of human experiences, particularly telling the stories of researchers and clinicians helping unravel the enduring mysteries of the human mind.

LINDA McIVER is the author of *Raising Heretics: Teaching kids to change the world,* and the founder and executive director of the Australian Data Science Education Institute charity. A passionate advocate for data literacy and engaging, empowering education for all, Linda also hosts the *Make Me Data Literate* podcast.

FIONA McMILLAN-WEBSTER is a science writer with a BSc in physics and a PhD in biophysics. Her writing has appeared in *Cosmos Magazine, Australian Geographic, National Geographic,*

Forbes, and others. Her first book, *The Age of Seeds: How plants hacked time and why our future depends on it*, was shortlisted for the 2023 Mark and Evette Moran Nib Literary Prize and went on to win the People's Choice Prize.

SALLY MONTGOMERY is a post-doctoral research associate in social anthropology at the University of Cambridge, where she completed her PhD in 2025. Sally is an environmental anthropologist with a current focus on Lord Howe Island. Her research explores notions of human and non-human belonging and the politics of conservation.

FELICITY NELSON is a freelance science journalist and the founder of a strategy consulting company called Frogs and Stars. She has bylines in *Nature*, *Veritasium*, ScienceAlert, *Chemistry & Engineering News*, *Guardian Australia* and ABC News. Her stories were published in *The Best Australian Science Writing* anthologies in 2020, 2019 and 2017.

ELLEN PHIDDIAN is an Adelaide-based science journalist with a background in science communication and a particular interest in chemistry and energy. She has worked in the *Cosmos Magazine* and ABC Science newsrooms. Her work was also published in *The Best Australian Science Writing 2024*.

JAMES PURTILL is the ABC's online technology reporter, covering stories from social media to solar panels, and from artificial intelligence to electric vehicles. He hosted the ABC podcast *Hello AI Overlords* that won the 2024 Eureka Award for Science Journalism. He lives in Perth.

DREW ROOKE is a journalist and the assistant science and technology editor at *The Conversation Australia + NZ*. He is the author *of One Last Spin: The power and peril of the pokies* (Scribe,

2018) and *A Witness of Fact: The peculiar case of Chief Forensic Pathologist Colin Manock* (Scribe, 2022), which was shortlisted for the 2022 Ned Kelly Award for Best True Crime Writing.

JACKSON RYAN is an award-winning science journalist from Adelaide and president of the Science Journalists Association of Australia. His work focuses on scientific integrity, research misconduct and trust in science. He co-edited the *The Best Australian Science Writing 2024* anthology (you should buy that one, too!).

BELINDA SMITH became a science journalist after realising she wasn't going to cut it as a researcher. She joined the ABC in 2017 and currently hosts the Radio National program/podcast *Lab Notes*, which delves into the science behind new discoveries and current events.

CARL SMITH is a Walkley Award–winning science journalist and co-editor of *The Best Australian Science Writing 2024*. He's been a Eureka Prize for Science Journalism finalist three times, jointly winning in 2021. He makes audio documentaries and features for ABC Radio National and co-hosts kids ethics podcast *Short & Curly*.

ALICIA SOMETIMES is a poet, writer and broadcaster. She has performed her poetry at many venues, festivals and events around the world. In 2023 she received ANAT's Synapse Artist Residency and co-created an art installation for Science Gallery Melbourne's exhibition 'Dark Matters'. Her new book is *Stellar Atmospheres*.

CLARE WATSON is a freelance science journalist with a background in biomedical science. Since trading her pipettes for a pen, her work has appeared in *Undark*, *Nature*, *Hakai Magazine*, *Cosmos Magazine*, and *Australian Geographic*. She writes and fact-checks for ScienceAlert, and often tells stories that reveal how science unfolds and who scientists really are.

SARA WEBB is an award-winning astrophysicist, AI researcher and science communicator at Swinburne University. She explores cosmic phenomena using machine learning and leads student experiments aboard the International Space Station. A Forbes 30 Under 30 honouree, her debut book, *The Little Book of Cosmic Catastrophes*, launched globally in 2024.

CAT WILLIAMS is a freelance science writer based in Naarm/ Melbourne with an interest in zoology, Aboriginal science and the nature–culture divide. Her work can be found in *Meanjin*, *Griffith Review*, *The Saturday Paper* and *ParticleWA*, among others. She was shortlisted for the Kat Muscat Fellowship in 2025.

ACKNOWLEDGEMENTS

'The night I accidentally became a corpse flower's bedside manservant' by Angus Dalton was originally published under the same title in the *Sydney Morning Herald* on 24 January 2025 and is available at <www.smh.com.au/national/nsw/the-night-i-accidentally-became-a-corpse-flower-s-bedside-manservant-20250124-p5l6wq.html>

'Humanoid robots: Made in our image' by Owen Cumming was originally published under the same title in *Cosmos Magazine* on 3 March 2025.

'Mathematical transformations: The power of vectors and tensors' by Robyn Arianrhod is from the prologue of *Vector: A surprising story of space, time, and mathematical transformation*, a book published by UNSW Press on 1 July 2024 and by University of Chicago Press on 29 May 2024.

'Solomon Islands tribes sell carbon credits, not their trees' by Jo Chandler was originally published under the same title in *Yale Environment 360* on 16 April 2024 and is available at <e360.yale.edu/features/solomon-islands-sirebe-carbon-credits>

'Cat-astrophe: Australia's feral cat problem' by Cat Williams was originally published under the same title in ParticleWA on 12 July 2024 and is available at <particle.scitech.org.au/people/cat-astrophe-australias-feral-cat-problem/>

'When did your kidney stone start growing? ANTSO scientist carbon dated his to find out' by Jacinta Bowler was originally published under the same title in ABC Science on 6 September 2024 and is available at <www.abc.net.au/news/science/2024-09-06/kidney-stones-carbon-dating-health/104217876>

'China's race for fusion energy' by Gemma Conroy was originally published as 'Inside China's race to lead the world in nuclear fusion' in *Nature* on 28 August 2024 and is available at <www.nature.com/articles/d41586-024-02759-x>

'"Earth poetry" in the Arctic' by Lucinda Duxbury was originally published under the same title in *Cosmos Magazine* in March 2025.

'Fishing for a glacier's secrets' by Mark Horstman was originally published under the same title on the Australian Antarctic Program Partnership website on 7 February 2025 and is available at <aapp.shorthandstories.com/fishing-for-a-glaciers-secrets/>

'Is this actually PTSD? Clinicians divided over redefining borderline personality disorder' by Natasha May was originally published under the same title in the *Guardian* on 12 May 2024 and is available at <www.theguardian.com/society/article/2024/may/12/is-this-actually-ptsd-clinicians-divided-over-redefining-borderline-personality-disorder>

'A freediver finds belonging without breath' was originally published on SAPIENS under a Creative Commons Attribution-No Derivatives 4.0 International licence on 2 May 2024. The original article is available at <www.sapiens.org/culture/freediving-belonging-kinship/>

'Faster higher stronger doper' by Mathew Ward Agius was originally published under the same title in *Cosmos Magazine* on 6 June 2024.

'Insect consciousness' by Amalyah Hart was originally published under the same title in *Cosmos Magazine* in September 2024.

'Air conditioning quietly changed Australian life in just a few decades' by James Purtill was originally published under the same title in ABC News on 28 January 2025 and is available at <www.abc.net.au/news/science/2025-01-28/air-conditioning-changed-australia-technology-heat-comfort/104741512>

'Mirror molecules: The twisted problem of chemical chirality' by Ellen Phiddian was originally published under the same title in *Cosmos Magazine* on 3 March 2025.

'How surviving a deadly animal encounter is not the end of the trauma' by Angela Heathcote was originally published under the same title on ABC News on 28 November 2024 and is available at <www.abc.net.au/news/2024-11-28/woman-stung-by-blue-ringed-octopus-shocked-by-victim-blaming/104372866>

'Some psychedelic medicine developers want to ditch the therapy aspect. What could go wrong?' by Rich Haridy was originally published under the same title in *Salon* on 7 June 2024 and is available at <www.salon.com/2024/06/07/some-psychedelic-medicine-developers-want-to-ditch-the-therapy-aspect-what-could-go-wrong/>

'A snowflake in hell' by Jackson Ryan was originally published under the same title in *The Monthly* in August 2024 and is available at <www.themonthly.com.au/issue/2024/august/jackson-ryan/snowflake-hell>

'Staying faithful to Earth: A meditation on our planetary kin in the universe' by Ceridwen Dovey was originally published as 'Staying faithful to Earth' in the *Griffith Review* on 1 November 2024.

'Whale talk' by Drew Rooke was originally published under the same title in *Cosmos Magazine* on 26 September 2024.

'Gaza: Why is it so hard to establish the death toll?' by Smriti Mallapaty was originally published under the same title in *Nature* on 24 September 2024.

'Sounds of the slow-rolling sea' by Sara Webb was originally published under the same title in *Cosmos Magazine* in March 2024.

'Jocelyn Bell Burnell, Mae Jemison, Carolyn Beatrice Parker' by Alicia Sometimes were originally published as three poems in *Stellar Atmospheres* by Cordite Press on 1 April 2024.

'Smooth fade' by Zowie Douglas-Kinghorn was originally published under the same title in *Meanjin Quarterly* on 2 December 2024.

'Why can't we remember our lives as babies or toddlers?' by Donna Lu was originally published under the same title in the *Guardian* on 16 February 2025 and is available at <www.theguardian.com/science/2025/feb/16/why-cant-we-remember-our-lives-as-babies-or-toddlers>

'Using trash to track other trash' by Clare Watson was originally published under the same title in *Hakai Magazine* on 6 June 2025 and is available at <hakaimagazine.com/news/using-trash-to-track-other-trash/>

'Alternative accommodation' by Anne Casey was originally published under the same title in *Live Canon International Prize Anthology* on 1 January 2025.

'Ethical problems continue to plague biometric studies of Chinese minority groups' by Dyani Lewis was originally published as 'Unethical studies on Chinese minority groups are being retracted – but not fast enough, critics say' in *Nature* on 24 January 2024 and is available at <www.nature.com/articles/d41586-024-00170-0>

'From hypnotised to heretic: Immunising society against misinformation' by Linda McIver was originally published under the same title on *ADSEI* on 23 November 2024 and is available at <adsei.org/2024/11/23/from-hypnotised-to-heretic-immunising-society-against-misinformation/>

'Skink on the brink' by Anthea Batsakis was originally published under the same title in *Australian Geographic* on 9 December 2024 and is available at <www.australiangeographic.com.au/topics/wildlife/2024/12/skink-on-the-brink/>

'My partner and I both have Long COVID. We tread the underworld together' by Felicity Nelson was originally published under the same title in the *Sick Times* on 25 June 2024.

'Moments of kindness in a regional hospital' by Michael Leach was originally published under the same title in *Kindness Anthology: Stories of kindness in healthcare* (Hush Foundation) on 21 November 2024.

'Mystery seismic waves that rippled around Earth for nine days caused by Greenland tsunami: Study' by Carl Smith and Peter de Kruijff was originally published under the same title on ABC News on 13 September 2024 and is available at <www.abc.net.au/news/science/2024-09-13/mystery-seismic-signal-greenland-landslide-tsunami-seiche-fjord/104327900>

'The world has been its hottest on record for ten months straight. Scientists can't fully explain why' by Tyne Logan was originally published under the same title on ABC News on 9 April 2024 and is available at <www.abc.net.au/news/2024-04-09/data-can-t-explain-off-the-charts-heat/103649190>

'Why some venomous snakes can bite and kill even when they're dead and decapitated' by Belinda Smith was originally published under the same title on ABC News on 2 June 2024 and is available at <www.abc.net.au/news/science/2024-06-02/snakes-deadly-bite-venom-dead-kill-what-the-duck/103913848>

'A museum heist 70 years ago is still causing a flutter in butterfly science today' by Olivia Congdon was originally published under the same title on the Australian National University College of Science and Medicine website on 19 February 2025 and is available at <science.anu.edu.au/news-events/news/museum-heist-70-years-ago-still-causing-flutter-butterfly-science-today>

'Stanford prison experiment: Why the lead psychologist defended his infamous study to the end' by Rachel Fieldhouse was originally published under the same name on AusDoc on 4 November 2024 and is available at <www.ausdoc.com.au/news/stanford-prison-researcher-dies-why-the-psychologist-defended-his-experiment-to-the-end/>

'Deep time encounters in the garden' by Fiona McMillan-Webster was originally published under the same title in *The Museum* in August 2024.

'The unexpected poetry of PhD acknowledgements' by Tabitha Carvan was published under the same title on the Australian National University College of Science and Medicine website on 11 July 2024 and is available at <science.anu.edu.au/news-events/news/unexpected-poetry-phd-acknowledgements>

2025 Shortlist

In 2012, UNSW Press launched an annual prize for the best short non-fiction piece on science written for a general audience. The UNSW Press Bragg Prize for Science Writing is named in honour of Australia's first Nobel laureates, William Henry Bragg and his son William Lawrence Bragg. The Braggs won the 1915 Nobel Prize for physics for their work on the analysis of crystal structure by means of X-rays. Both scientists led enormously productive lives and left a lasting legacy. William Henry Bragg was a firm believer in making science popular among young people, and his lectures for students were described as models of clarity and intellectual excitement.

The UNSW Press Bragg Prize is supported by the Copyright Agency Cultural Fund and UNSW Science. The winner receives a prize of $7000 and two runners-up each receive a prize of $1500.

The shortlisted entries for the 2025 prize are included in this anthology.

The UNSW Press Bragg Prize for Science Writing 2025 Shortlist

Tabitha Carvan, 'The unexpected poetry of PhD acknowledgements'

Angus Dalton, 'The night I accidentally became a corpse flower's bedside manservant'

Lucinda Duxbury, '"Earth poetry" in the Arctic'

Linda McIver, 'From hypnotised to heretic: Immunising society against misinformation'

Sally Montgomery, 'A freediver finds belonging without breath'

James Purtill, Air conditioning quietly changed Australian life in just a few decades'

Winners announced in November 2025.
<unsw.press/the-unsw-press-bragg-prize-for-science-writing/>

Judges for the UNSW Press Bragg Prize 2025

Professor Merlin Crossley, University of New South Wales
Professor Joan Leach, Australian National University
Emeritus Professor Ian Lowe, Griffith University
Zoe Kean, co-editor, *The Best Australian Science Writing 2025*
Tegan Taylor, co-editor, *The Best Australian Science Writing 2025*

The Bragg Prize hosted a special category for high school students between 2015 and 2024. Science enthusiasts in years 7 to 10 were invited to submit an essay of up to 800 words in a joint initiative of UNSW Press, UNSW Science and Refraction Media, with support from the Copyright Agency Cultural Fund.

The 2024 competition asked students to write a short essay on 'People power: Working together to protect our environment. Biodiversity, community and conservation?' Entries in 2024 were judged by a panel comprising Donna Buckley, mathematics and cybersecurity teacher at John Curtin College of the Arts; Corey Tutt, CEO and founder of DeadlyScience; wildlife biologist Janice Vas; and Jackson Ryan and Carl Smith, co-editors of *The Best Australian Science Writing 2024*.

The winning entry was 'Biodiversity and community: Butterfly counts' by Reena Du, a year 10 student from Abbotsleigh, Wahroonga, NSW. Her winning essay is included on the following pages.

The UNSW Press Bragg Student Prize for Science Writing
2024 Winner

BIODIVERSITY AND COMMUNITY: BUTTERFLY COUNTS

Reena Du
Year 10, Abbotsleigh, Wahroonga, NSW

Imagine being in a sunny garden, the air filled with the soft flutter of wings, each a different species of butterfly. This is what volunteers at Brisbane's Big Butterfly Count experience, where citizens are able to see what it is like as a scientist to document the gentle beauty of butterflies. By carefully tracking each butterfly's presence and living patterns, butterfly counts provide a window into the health of our environment, allowing us to foresee and reveal shifts in biodiversity and ecological balances. This blend of scientific research and natural beauty allows every butterfly to be turned into data, giving way to the connections between them and the ecosystems they inhabit.

Brisbane's Big Butterfly Count started in the butterfly season of 2020–21, created from an initiative by the Brisbane Catchments Network. The project aims to engage residents of Brisbane of all ages and education levels to experience and contribute to a greater project of understanding our ecosystems, with around 160 different butterfly species being recorded. This annual event not only enhances the understanding of local butterfly populations but also exhibits the power of citizen scientists in an urban setting.

But why do we count butterflies?

Butterflies play a vital role in the ecosystem as pollinators and being a component of the food chain allows them to be indicators of environmental change. Their sensitivity to changes and quick reactions in their habitats make them valuable for monitoring shifts in ecological conditions.

Butterflies face many challenges that threaten their survival. They experience a loss of habitat, which can be linked to urban development, agricultural trends and deforestation. These activities reduce the areas butterflies use for feeding, breeding and sheltering. Climate change worsens these issues by altering weather patterns, disrupting the lifecycle of butterflies, such as times of eclosing, and the availability of food sources. Additionally, the use of pesticides and herbicides in crop growing poses a threat to butterflies as they can potentially kill them and harm the plants they collect pollen and nectar from. Invasive species can disrupt the lives of butterflies by outcompeting with them to plants, and could possibly prey on butterflies and their larvae.

By taking part in butterfly counts, community members provide data that will be able to help scientists and researchers track changes in the environment, allowing them to make informed decisions about conservational strategies.

However, not only do butterfly counts help the environment, research shows that they also calm humans. They give individuals a sense of connection to nature, and evidence shows that an increase in this connection leads to an increase in human wellbeing and pro-environmental behaviours. A study was conducted by researcher Carly Butler on emotions participants felt during a butterfly count, and it was shown that the more emotions felt, the more engagement with nature.

What does Brisbane's Big Butterfly Count do?

Brisbane's Big Butterfly Count provides many opportunities for researchers and butterfly enthusiasts to engage with butterflies and the ecosystems they inhabit.

Surveys are conducted at a variety of locations in Brisbane, representing different butterfly habitats. The data is collected three times during the butterfly season. Citizens use the BioCollect app or paper flyers featuring over 30 common local butterflies, allowing easy accessibility for people who may not be used to research. The data is entered into the Atlas of Living Australia to keep track of the butterflies found.

There are many workshops for those wanting to learn more about butterflies, such as how to identify butterflies and the areas they inhabit. Ecology walks are also provided to explore the relationship between butterflies and their habitats. Educators and butterfly enthusiasts are given resources to spread their teachings to other people. Additionally, targeted butterfly plantings for butterfly rehabilitators and gardeners promote teachings on larval host plants. There are many public events highlighting the environmental value of looking after butterflies.

Overall, Brisbane's Big Butterfly Count is a great way for citizen scientists to observe the environment around them, while enjoying the butterflies. It not only fosters a deeper connection with the surroundings but also provides data for understanding and preserving biodiversity. These counts offer a window into the health of the environment, revealing insights on the impact humans have on the natural world.

ADVISORY PANEL

MERLIN CROSSLEY AM is Deputy Vice-Chancellor Academic Quality at UNSW. He studies CRISPR gene editing and blood diseases. Crossley is an enthusiastic science communicator, Chair of *The Conversation*'s Editorial Board and Deputy Director of the Australian Science Media Centre. He was a Rhodes Scholar, a recipient of the NSW Premier's Award for Medical Biological Science, and in 2021 a species of iridescent squid was named in his honour – *Iridoteuthis merlini* – Merlin's butterfly bobtail squid.

JOAN LEACH is Deputy Vice-Chancellor Academic at the ANU. Professor Leach was Director of the Australian National Centre for the Public Understanding of Science for ten years, until 2024, and has been president of Australian Science Communicators and chair of the National Committee for History and Philosophy of Science at the Australian Academy of Science. Her research centres on public engagement with science, medicine and technology, and she has published extensively about science communication, including the recent volume *An Ethics of Science Communication* (with Fabien Medvecky).

IAN LOWE is Emeritus Professor of Science, Technology and Society at Griffith University, where he was previously Head of the School of Science. He has published widely and served on many advisory bodies to all levels of government. He has received many awards for his work, notably being made an Officer of the Order of Australia in 2001. He has been a Fellow of the Academy of Technological Sciences and Engineering for twenty years.

www.ingramcontent.com/pod-product-compliance
Lightning Source LLC
Chambersburg PA
CBHW021458210326
41599CB00012B/1047